Teacher Edition

Eureka Math
Grade 3
Module 4

Special thanks go to the Gordon A. Cain Center and to the Department of Mathematics at Louisiana State University for their support in the development of *Eureka Math*.

For a free *Eureka Math* Teacher
Resource Pack, Parent Tip
Sheets, and more please
visit www.Eureka.tools

Published by the non-profit Great Minds

Printed in the U.S.A.
This book may be purchased from the publisher at eureka-math.org
10 9
ISBN 978-1-63255-366-9

Eureka Math: A Story of Units **Contributors**

Katrina Abdussalaam, Curriculum Writer
Tiah Alphonso, Program Manager—Curriculum Production
Kelly Alsup, Lead Writer / Editor, Grade 4
Catriona Anderson, Program Manager—Implementation Support
Debbie Andorka-Aceves, Curriculum Writer
Eric Angel, Curriculum Writer
Leslie Arceneaux, Lead Writer / Editor, Grade 5
Kate McGill Austin, Lead Writer / Editor, Grades PreK–K
Adam Baker, Lead Writer / Editor, Grade 5
Scott Baldridge, Lead Mathematician and Lead Curriculum Writer
Beth Barnes, Curriculum Writer
Bonnie Bergstresser, Math Auditor
Bill Davidson, Fluency Specialist
Jill Diniz, Program Director
Nancy Diorio, Curriculum Writer
Nancy Doorey, Assessment Advisor
Lacy Endo-Peery, Lead Writer / Editor, Grades PreK–K
Ana Estela, Curriculum Writer
Lessa Faltermann, Math Auditor
Janice Fan, Curriculum Writer
Ellen Fort, Math Auditor
Peggy Golden, Curriculum Writer
Maria Gomes, Pre-Kindergarten Practitioner
Pam Goodner, Curriculum Writer
Greg Gorman, Curriculum Writer
Melanie Gutierrez, Curriculum Writer
Bob Hollister, Math Auditor
Kelley Isinger, Curriculum Writer
Nuhad Jamal, Curriculum Writer
Mary Jones, Lead Writer / Editor, Grade 4
Halle Kananak, Curriculum Writer
Susan Lee, Lead Writer / Editor, Grade 3
Jennifer Loftin, Program Manager—Professional Development
Soo Jin Lu, Curriculum Writer
Nell McAnelly, Project Director

Ben McCarty, Lead Mathematician / Editor, PreK–5
Stacie McClintock, Document Production Manager
Cristina Metcalf, Lead Writer / Editor, Grade 3
Susan Midlarsky, Curriculum Writer
Pat Mohr, Curriculum Writer
Sarah Oyler, Document Coordinator
Victoria Peacock, Curriculum Writer
Jenny Petrosino, Curriculum Writer
Terrie Poehl, Math Auditor
Robin Ramos, Lead Curriculum Writer / Editor, PreK–5
Kristen Riedel, Math Audit Team Lead
Cecilia Rudzitis, Curriculum Writer
Tricia Salerno, Curriculum Writer
Chris Sarlo, Curriculum Writer
Ann Rose Sentoro, Curriculum Writer
Colleen Sheeron, Lead Writer / Editor, Grade 2
Gail Smith, Curriculum Writer
Shelley Snow, Curriculum Writer
Robyn Sorenson, Math Auditor
Kelly Spinks, Curriculum Writer
Marianne Strayton, Lead Writer / Editor, Grade 1
Theresa Streeter, Math Auditor
Lily Talcott, Curriculum Writer
Kevin Tougher, Curriculum Writer
Saffron VanGalder, Lead Writer / Editor, Grade 3
Lisa Watts-Lawton, Lead Writer / Editor, Grade 2
Erin Wheeler, Curriculum Writer
MaryJo Wieland, Curriculum Writer
Allison Witcraft, Math Auditor
Jessa Woods, Curriculum Writer
Hae Jung Yang, Lead Writer / Editor, Grade 1

Mathematics Curriculum

3
GRADE

Table of Contents

GRADE 3 • MODULE 4

Multiplication and Area

Grade 3 • Module 4

Multiplication and Area

OVERVIEW

In this 20-day module, students explore area as an attribute of two-dimensional figures and relate it to their prior understandings of multiplication. In Grade 2, students partitioned a rectangle into rows and columns of same-sized squares and found the total number by both counting and adding equal addends represented by the rows or columns (**2.G.2**, **2.OA.4**).

In Topic A, students begin to conceptualize area as the amount of two-dimensional surface that is contained within a plane figure. They come to understand that the space can be tiled with unit squares without gaps or overlaps (**3.MD.5**). Students decompose paper strips into square inches and square centimeters, which they use to tile 3 by 4, 4 by 3, and 2 by 6 rectangles. They compare rectangles tiled with like units and notice different side lengths but equal areas. Topic A provides students' first experience with tiling from which they learn to distinguish between length and area by placing a ruler with the same size units (inches or centimeters) next to a tiled array. They discover that the number of tiles along a side corresponds to the length of the side (**3.MD.6**).

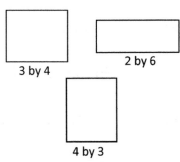

In Topic B, students progress from using square tile manipulatives to drawing their own area models. Anticipating the final structure of an array, they complete rows and columns in figures such as the example shown to the right. Students connect their extensive work with rectangular arrays and multiplication to eventually discover the area formula for a rectangle, which is formally introduced in Grade 4 (**3.MD.7a**).

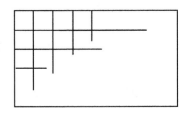

In Topic C, students manipulate rectangular arrays to concretely demonstrate the arithmetic properties in anticipation of the lessons that follow. They do this by cutting rectangular grids and rearranging the parts into new wholes using the properties to validate that area stays the same, despite the new dimensions. They apply tiling and multiplication skills to determine all whole number possibilities for the side lengths of rectangles given their areas (**3.MD.7b**).

Topic D creates an opportunity for students to solve problems involving area (**3.MD.7b**). Students decompose or compose composite regions, such as the one shown to the right—into non-overlapping rectangles, find the area of each region, and then add or subtract to determine the total area of the original shape. This leads students to find the areas of rooms in a given floor plan (**3.MD.7d**).

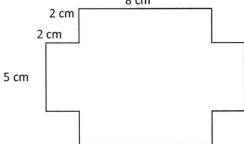

EUREKA
MATH™

Notes on Pacing for Differentiation

If pacing is a challenge, consider the following modifications and omissions.

Consolidate Lessons 2 and 3, both of which involve measuring and comparing area. From Lesson 3, omit the use of square centimeter tiles and the Application Problem. Have students establish square inches as units, and then tile with them as a strategy to measure area.

Consider omitting Lesson 9, which reviews previously learned skills. If omitting, be sure that students are ready to transition toward more complex practice.

Omit Lessons 15 and 16. These lessons guide students through a project involving floor plans. Skip the application of area that these lessons provide.

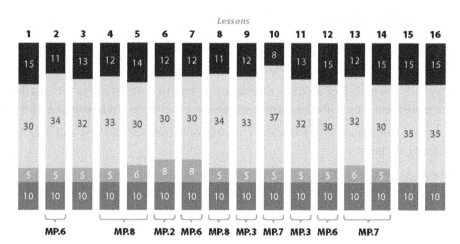

Focus Grade Level Standards

Geometric measurement: understand concepts of area and relate area to multiplication and to addition.

3.MD.5   Recognize area as an attribute of plane figures and understand concepts of area measurement.

a.   A square with side length 1 unit, called "a unit square," is said to have "one square unit" of area, and can be used to measure area.

b. A plane figure which can be covered without gaps or overlaps by *n* unit squares is said to have an area of *n* square units.

3.MD.6   Measure areas by counting unit squares (square cm, square m, square in, square ft, and improvised units).

3.MD.7   Relate area to the operations of multiplication and addition.

a. Find the area of a rectangle with whole-number side lengths by tiling it, and show that the area is the same as would be found by multiplying the side lengths.

b. Multiply side lengths to find areas of rectangles with whole-number side lengths in the context of solving real world and mathematical problems, and represent whole-number products as rectangular areas in mathematical reasoning.

c. Use tiling to show in a concrete case that the area of a rectangle with whole-number side lengths *a* and *b* + *c* is the sum of *a* × *b* and *a* × *c*. Use area models to represent the distributive property in mathematical reasoning.

d. Recognize area as additive. Find the areas of rectilinear figures by decomposing them into non-overlapping rectangles and adding the areas of the non-overlapping parts, applying this technique to solve real world problems.

Foundational Standards

2.MD.1   Measure the length of an object by selecting and using appropriate tools such as rulers, yardsticks, meter sticks, and measuring tapes.

2.MD.2   Measure the length of an object twice, using length units of different lengths for the two measurements; describe how the two measurements relate to the size of the unit chosen.

2.G.2   Partition a rectangle into rows and columns of same-size squares and count to find the total number of them.

2.OA.4   Use addition to find the total number of objects arranged in rectangular arrays with up to 5 rows and up to 5 columns; write an equation to express the total as a sum of equal addends.

Focus Standards for Mathematical Practice

MP.2   **Reason abstractly and quantitatively**. Students build toward abstraction, starting with tiling a rectangle, and then gradually move to finish incomplete grids and drawing grids of their own. Students then eventually work purely in the abstract, imagining the grid as needed.

MP.3   **Construct viable arguments and critique the reasoning of others**. Students explore their conjectures about area by cutting to decompose rectangles and then recomposing them in different ways to determine if different rectangles have the same area. When solving area problems, students learn to justify their reasoning and determine whether they have found all possible solutions, when multiple solutions are possible.

MP.6   **Attend to precision**. Students precisely label models and interpret them, recognizing that the unit impacts the amount of space a particular model represents, even though pictures may appear to show equal-sized models. They understand why, when side lengths are multiplied, the result is given in square units.

MP.7 **Look for and make use of structure.** Students relate previous knowledge of the commutative and distributive properties to area models. They build from spatial structuring to understanding the number of area-units as the product of number of units in a row and number of rows.

MP.8 **Look for and express regularity in repeated reasoning.** Students use increasingly sophisticated strategies to determine area throughout the course of the module. As students analyze and compare strategies, they eventually realize that area can be found by multiplying the number in each row by the number of rows.

Overview of Module Topics and Lesson Objectives

Standards		Topics and Objectives	Days
3.MD.5 **3.MD.6** **3.MD.7**	A	**Foundations for Understanding Area**	4
		Lesson 1: Understand area as an attribute of plane figures.	
		Lesson 2: Decompose and recompose shapes to compare areas.	
		Lesson 3: Model tiling with centimeter and inch unit squares as a strategy to measure area.	
		Lesson 4: Relate side lengths with the number of tiles on a side.	
3.MD.5 **3.MD.6** **3.MD.7a** **3.MD.7b** **3.MD.7d**	B	**Concepts of Area Measurement**	4
		Lesson 5: Form rectangles by tiling with unit squares to make arrays.	
		Lesson 6: Draw rows and columns to determine the area of a rectangle given an incomplete array.	
		Lesson 7: Interpret area models to form rectangular arrays.	
		Lesson 8: Find the area of a rectangle through multiplication of the side lengths.	
		Mid-Module Assessment: Topics A–B (assessment 1 day, return ½ day, remediation or further applications ½ day)	2
3.MD.5 **3.MD.7a** **3.MD.7b** **3.MD.7c** **3.MD.7d**	C	**Arithmetic Properties Using Area Models**	3
		Lesson 9: Analyze different rectangles and reason about their area.	
		Lesson 10: Apply the distributive property as a strategy to find the total area of a large rectangle by adding two products.	
		Lesson 11: Demonstrate the possible whole number side lengths of rectangles with areas of 24, 36, 48, or 72 square units using the associative property.	

Standards	Topics and Objectives		Days
3.MD.6 3.MD.7a 3.MD.7b 3.MD.7c 3.MD.7d 3.MD.5	D	**Applications of Area Using Side Lengths of Figures** Lesson 12: Solve word problems involving area. Lessons 13–14: Find areas by decomposing into rectangles or completing composite figures to form rectangles. Lessons 15–16: Apply knowledge of area to determine areas of rooms in a given floor plan.	5
	End-of-Module Assessment: Topics A–D (assessment 1 day, return ½ day, remediation or further applications ½ day)		2
Total Number of Instructional Days			**20**

Terminology

New or Recently Introduced Terms

- Area (the amount of two-dimensional space in a bounded region)
- Area model (a model for multiplication that relates rectangular arrays to area)

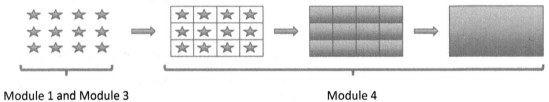

Module 1 and Module 3   Module 4

- Square unit (a unit of area—specifically square centimeters, inches, feet, and meters)
- Tile (to cover a region without gaps or overlaps)
- Unit square (e.g., given a length unit, it is a 1 unit by 1 unit square)
- Whole number (an integer, i.e., a number without fractions)

Familiar Terms and Symbols[1]

- Array (a set of numbers or objects that follow a specific pattern: a matrix)
- Commutative property (e.g., rotate a rectangular array 90 degrees to demonstrate that factors in a multiplication sentence can switch places)
- Distribute (e.g., $2 \times (3 + 4) = 2 \times 3 + 2 \times 4$)
- Geometric shape (a two-dimensional object with a specific outline or form)
- Length (the straight-line distance between two points)

---

[1]These are terms and symbols students have seen previously.

EUREKA
MATH™

- Multiplication (e.g., 5 × 3 = 15)
- Rows and columns (e.g., in reference to rectangular arrays)

Suggested Tools and Representations

- Area model
- Array
- Grid paper (inch and centimeter)
- Rulers (both centimeter and inch measurements)
- Unit squares in both inch and centimeter lengths (e.g., square tiles used for measuring area—can be made out of paper if plastic or wood tiles are not available)

Scaffolds[2]

The scaffolds integrated into *A Story of Units* give alternatives for how students access information as well as express and demonstrate their learning. Strategically placed margin notes are provided within each lesson elaborating on the use of specific scaffolds at applicable times. They address many needs presented by English language learners, students with disabilities, students performing above grade level, and students performing below grade level. Many of the suggestions are organized by Universal Design for Learning (UDL) principles and are applicable to more than one population. To read more about the approach to differentiated instruction in *A Story of Units,* please refer to "How to Implement *A Story of Units.*"

Assessment Summary

Type	Administered	Format	Standards Addressed
Mid-Module Assessment Task	After Topic B	Constructed response with rubric	3.MD.5 3.MD.6 3.MD.7abd
End-of-Module Assessment Task	After Topic D	Constructed response with rubric	3.MD.5 3.MD.6 3.MD.7a–d

---

[2]Students with disabilities may require Braille, large print, audio, or special digital files. Please visit the website, www.p12.nysed.gov/specialed/aim, for specific information on how to obtain student materials that satisfy the National Instructional Materials Accessibility Standard (NIMAS) format.

EUREKA
MATH™

Module 4:   Multiplication and Area

©2015 Great Minds. eureka math.org
G3-M4-TE-B4-1.3.1-01.2016   -

Mathematics Curriculum

3
GRADE

Topic A

Foundations for Understanding Area

3.MD.5, 3.MD.6, 3.MD.7

Focus Standard:	3.MD.5	Recognize area as an attribute of plane figures and understand concepts of area measurement:
		a. A square with side length 1 unit, called "a unit square," is said to have "one square unit" of area, and can be used to measure area.
		b. A plane figure which can be covered without gaps or overlaps by *n* unit squares is said to have an area of *n* square units.
Instructional Days:	4	
Coherence -Links from:	G2–M2	Addition and Subtraction of Length Units
	G3–M1	Properties of Multiplication and Division and Solving Problems with Units of 2–5 and 10
	G3–M3	Multiplication and Division with Units of 0, 1, 6–9, and Multiples of 10
-Links to:	G4–M3	Multi-Digit Multiplication and Division
	G4–M7	Exploring Multiplication

In Lesson 1, students come to understand area as an attribute of plane figures that is defined by the amount of two-dimensional space within a bounded region. Students use pattern blocks to tile given polygons without gaps or overlaps and without exceeding the boundaries of the shape.

Lesson 2 takes students into an exploration in which they cut apart paper rectangles into same-size squares to concretely define a square unit, specifically square inches and centimeters. They use these units to make rectangular arrays that have the same area but different side lengths.

Lessons 3 and 4 introduce students to the strategy of using centimeter and inch tiles to find area. Students use tiles to determine the area of a rectangle by tiling the region without gaps or overlaps. They then bring the ruler (with corresponding units) alongside the array to discover that the side length is equal to the number of tiles required to cover one side of the rectangle. From this experience, students begin relating total area with multiplication of side lengths.

A Teaching Sequence Toward Mastery of Foundations for Understanding Area

Objective 1: Understand area as an attribute of plane figures.
(Lesson 1)

Objective 2: Decompose and recompose shapes to compare areas.
(Lesson 2)

Objective 3: Model tiling with centimeter and inch unit squares as a strategy to measure area.
(Lesson 3)

Objective 4: Relate side lengths with the number of tiles on a side.
(Lesson 4)

Lesson 1

Objective: Understand area as an attribute of plane figures.

Suggested Lesson Structure

■ Fluency Practice   (15 minutes)
■ Application Problem   (5 minutes)
□ Concept Development   (30 minutes)
■ Student Debrief   (10 minutes)

   Total Time   **(60 minutes)**

Fluency Practice (15 minutes)

- Group Counting **3.OA.1**   (4 minutes)
- Identify the Shape **2.G.1**   (3 minutes)
- Find the Common Products **3.OA.7**   (8 minutes)

Group Counting (4 minutes)

Note: Group counting reviews interpreting multiplication as repeated addition.

Instruct students to count forward and backward, occasionally changing the direction of the count.

- Threes to 30
- Sixes to 60
- Sevens to 70
- Eights to 80
- Nines to 90

Identify the Shape (3 minutes)

Note: This fluency activity reviews properties and vocabulary that are used during today's Concept Development.

T: (Project a triangle.) How many sides does this shape have?
S: 3.
T: Name the shape.
S: Triangle.

Continue with the following possible sequence: quadrilateral (trapezoid), quadrilateral (rhombus), quadrilateral (square), and quadrilateral (rectangle).

EUREKA
MATH™

Find the Common Products (8 minutes)

Materials: (S) Blank paper

Note: This fluency activity reviews multiplication patterns from Module 3.

T: Fold your paper in half vertically. Unfold your paper. On the left half, count by twos to 20 down the side of your paper. On the right half, count by fours to 40 down the side of your paper. Draw lines to match products that appear in both columns.

S: (Match 4, 8, 12, 16, and 20.)

T: (Write ___ × 2 = 4,   ___ × 2 = 8, etc., next to each corresponding product on the left half of the paper.) Write the complete equations next to their products.

S: (Write equations and complete unknowns.)

T: (Write 4 = ____ × 4, 8 = ____ × 4, etc., next to each corresponding product on the right half of the paper.) Write the complete equations next to their products.

S: (Write equations.)

T: (Write 2 × 2 = ____ × 4.) Say the equation, including all factors.

S: 2 × 2 = 1 × 4.

T: (Write 2 × 2 = 1 × 4.) Write the remaining equal facts as equations.

S: (Write 4 × 2 = 2 × 4, 6 × 2 = 3 × 4, 8 × 2 = 4 × 4, and 10 × 2 = 5 × 4.)

T: What patterns do you notice in your equations?

S: Each multiple of 4 is also a multiple of 2.

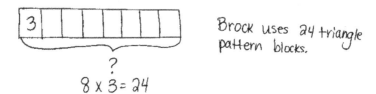

Application Problem (5 minutes)

Eric makes a shape with 8 trapezoid pattern blocks. Brock makes the same shape using triangle pattern blocks. It takes 3 triangles to make 1 trapezoid. How many triangle pattern blocks does Brock use?

Note: This problem reviews the Module 3 concept of multiplying using units of 8.

Concept Development (30 minutes)

Materials: (S) Pattern blocks, Problem Set

Part 1: Use pattern blocks to understand area.

NOTES ON
MULTIPLE MEANS
OF ACTION AND
EXPRESSION:

Manipulating pattern blocks may be a
challenge for some learners. Try the
following tips:

- Partner students so they can work
  together to cover the shapes.

- Encourage students to hold the
  pattern blocks in place with one
  hand, while they place the remaining
  blocks.

- Instead of using pattern blocks,
  provide paper shapes that can be
  glued, so they will not move around
  unnecessarily.

- Offer the computer as a resource to
  create and move shapes.

T: Look at Problem 1 on your Problem Set. Discuss with a
partner whether you think Shape A or B takes up more
space. Be prepared to explain your answer. (After
students discuss, facilitate a whole class discussion.)

S: Shape A because it's longer than Shape B. → Shape B
because it's taller than Shape A.

T: Use triangle pattern blocks to cover Shapes A and B.
Be sure the triangles do not have gaps between them,
do not overlap, and do not go outside the sides of the
shapes. (Allow time for students to work.) What did
you notice about the number of triangles it takes to
cover Shapes A and B?

S: It takes 6 triangles to cover each shape!

T: Answer Problem 1 on your Problem Set. (Allow time
for students to write answers.)
Do all the triangles you used to cover Shapes A and B take up the same amount of space?

S: Yes because they're all the same size.

T: What does that mean about the amount of space taken up by Shapes A and B?

S: They are the same. → It took 6 triangles to cover each shape, which means the shapes take up the
same amount of space. → The amount of space that Shape A takes up is equal to the amount of
space taken up by Shape B.

T: The amount of flat space a shape takes up is called its **area**. Because Shapes A and B take up the
same amount of space, their areas are equal.

Repeat the process of using pattern blocks to cover Shapes A and B with the rhombus and trapezoid pattern
blocks. Students record their work on Problems 2 and 3 in the Problem Set.

T: What is the relationship between the size of the pattern blocks and the number of pattern blocks it
requires to cover Shapes A and B?

S: The bigger the pattern block, the smaller the number of pattern blocks it requires to cover these
shapes. → The bigger pattern blocks, like the trapezoid, cover more area than the triangles. That
means it takes fewer trapezoids to cover the same area as the triangles.

T: Answer Problem 4 on your Problem Set.

Part 2: Measure area using square units.

T: Use square pattern blocks to cover the rectangle in Problem 5. Be sure the squares do not have gaps between them, do not overlap, and do not go outside the sides of the rectangle. (Allow students time to work.) How many squares did you need to cover the rectangle?

S: 6.

T: Answer Problem 5 on your Problem Set. (Allow time for students to write answers.) The area of the rectangle is 6 **square units**. Why do you think we call them square units?

S: Because they are squares! → The units used to measure are squares, so they are square units!

T: Yes! The units used to measure the area of the rectangle are squares.

T: Use trapezoid pattern blocks to cover the rectangle in Problem 5. Be sure the trapezoids do not have gaps between them, do not overlap, and do not go outside the sides of the rectangle. (Allow students time to work.) What did you notice?

S: It's not possible! → The trapezoids cannot cover this shape without having gaps, overlapping, or going outside the lines.

T: Use this information to help you answer Problem 6 on your Problem Set. (Allow time for students to write answers.) I'm going to say an area in square units, and you are going to make a rectangle with your pattern blocks having that area. Which pattern blocks will you use?

S: The squares because the units for area that you are telling us are square units!

T: Here we go! Four square units.

S: (Make rectangles.)

Continue with the following possible sequence: 12 square units, 9 square units, and 8 square units. Invite students to compare their rectangles to a partner's rectangles. How are they the same? How are they different? If time allows, students can work with a partner to create rectangles that have the same areas, but look different.

> **NOTES ON MULTIPLE MEANS OF ENGAGEMENT:**
>
> Students working above grade level can be encouraged to find other square units in the classroom that they can either use to make rectangles or that already form rectangles. Such items might include sticky notes, desktops, floor tiles, and linking cubes. Students can create a poster to share with the class that shows the areas of the rectangles made with these other square units.

Student Debrief (10 minutes)

Lesson Objective: Understand area as an attribute of plane figures.

The Student Debrief is intended to invite reflection and active processing of the total lesson experience.

Invite students to review their solutions for the Problem Set. They should check work by comparing answers with a partner before going over answers as a class. Look for misconceptions or misunderstandings that can be addressed in the Debrief. Guide students in a conversation to debrief the Problem Set and process the lesson.

Any combination of the questions below may be used to lead the discussion.

- Talk to a partner. Do you think you can use square pattern blocks to cover Shapes A and B in Problem 1? Explain your answer.

- How many triangle pattern blocks does it take to cover a rhombus pattern block? Use that information to say a division fact that relates the number of triangles it takes to cover Shape A to the number of rhombuses it takes to cover the same shape. (6 ÷ 2 = 3.)

- Explain to a partner how you used square pattern blocks to find the area of the rectangle in Problem 5.

- What new math vocabulary did we use today to communicate precisely about the amount of space taken up by a shape? (**Area**.) Which units did we use to measure area? (**Square units**.)

- How did the Application Problem connect to today's lesson?

Exit Ticket (3 minutes)

After the Student Debrief, instruct students to complete the Exit Ticket. A review of their work will help with assessing students' understanding of the concepts that were presented in today's lesson and planning more effectively for future lessons. The questions may be read aloud to the students.

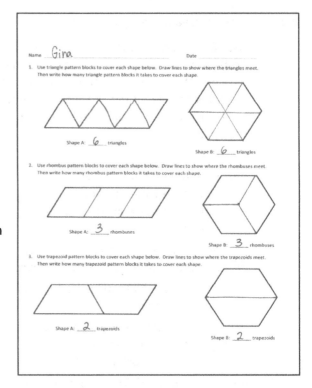

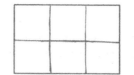

Lesson 1:   Understand area as an attribute of plane figures.

©2015 Great Minds. eureka-math.org
G3-M4-TE-B4-1.3.1-01.2016

EUREKA MATH

Name _____   Date _____

1. Use triangle pattern blocks to cover each shape below. Draw lines to show where the triangles meet. Then, write how many triangle pattern blocks it takes to cover each shape.

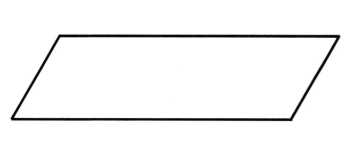

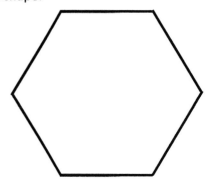

Shape A: _____ triangles

Shape B: _____ triangles

2. Use rhombus pattern blocks to cover each shape below. Draw lines to show where the rhombuses meet. Then, write how many rhombus pattern blocks it takes to cover each shape.

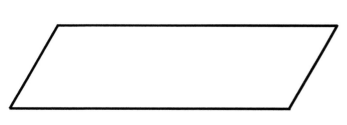

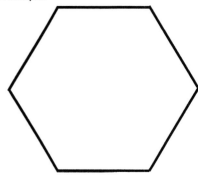

Shape A: _____ rhombuses

Shape B: _____ rhombuses

3. Use trapezoid pattern blocks to cover each shape below. Draw lines to show where the trapezoids meet. Then, write how many trapezoid pattern blocks it requires to cover each shape.

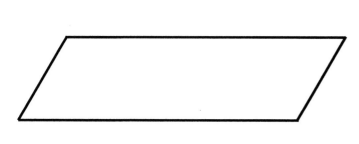

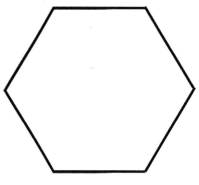

Shape A: _____ trapezoids

Shape B: _____ trapezoids

Lesson 1:   Understand area as an attribute of plane figures.

15

©2015 Great Minds. eureka math.org
G3-M4-TE-B4-1.3.1-01.2016 -

4.   How is the number of pattern blocks needed to cover the same shape related to the size of the pattern blocks?

5.   Use square pattern blocks to cover the rectangle below. Draw lines to show where the squares meet. Then, write how many square pattern blocks it requires to cover the rectangle.

_____ squares

6.   Use trapezoid pattern blocks to cover the rectangle in Problem 5. Can you use trapezoid pattern blocks to measure the area of this rectangle? Explain your answer.

Lesson 1:   Understand area as an attribute of plane figures.

©2015 Great Minds. eureka-math.org
G3-M4-TE-B4-1.3.1-01.2016

EUREKA
MATH

Name _____   Date _____

Each ▢ is 1 square unit. Do both rectangles have the same area? Explain how you know.

Lesson 1:   Understand area as an attribute of plane figures.

©2015 Great Minds. eureka math.org
G3-M4-TE-B4-1.3.1-01.2016   -

17

Name _____   Date _____

1. Magnus covers the same shape with triangles, rhombuses, and trapezoids.

   a.   How many **triangles** will it take to cover the shape?

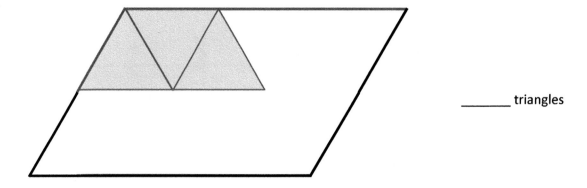

   _____ triangles

   b.   How many rhombuses will it take to cover the shape?

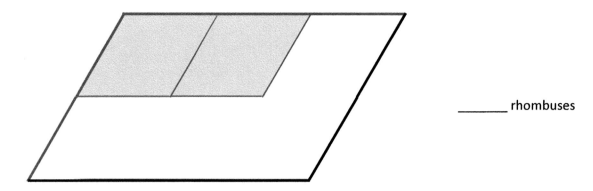

   _____ rhombuses

   c.   Magnus notices that 3 triangles from Part (a) cover 1 trapezoid. How many trapezoids will you need to cover the shape below? Explain your answer.

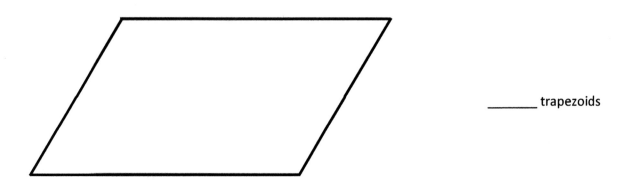

   _____ trapezoids

EUREKA
MATH™

2. Angela uses squares to find the area of a rectangle. Her work is shown below.

    a. How many squares did she use to cover the rectangle?

    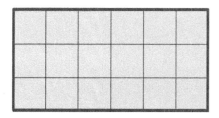

    _____ squares

    b. What is the area of the rectangle in square units? Explain how you found your answer.

3. Each 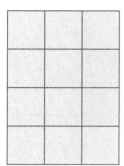 is 1 square unit. Which rectangle has the largest area? How do you know?

    Rectangle A

    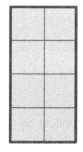

    Rectangle B

    Rectangle C

Lesson 2

Objective: Decompose and recompose shapes to compare areas.

Suggested Lesson Structure

■ Fluency Practice   (11 minutes)
■ Application Problem   (5 minutes)
■ Concept Development   (34 minutes)
■ Student Debrief   (10 minutes)
   Total Time   **(60 minutes)**

Fluency Practice (11 minutes)

- Group Counting **3.OA.1**   (4 minutes)
- Multiply by 4 **3.OA.7**   (7 minutes)

Group Counting (4 minutes)

Note: Group counting reviews interpreting multiplication as repeated addition.

Instruct students to count forward and backward, occasionally changing the direction of the count.

- Sixes to 60
- Sevens to 70
- Eights to 80
- Nines to 90

Multiply by 4 (7 minutes)

Materials: (S) Multiply by 4 (6–10) Pattern Sheet

Note: This activity builds fluency with respect to multiplication facts using units of 4. It works toward students knowing from memory all products of two one-digit numbers.

- T: (Write 7 × 4.) Let's skip-count up by fours. (Count with fingers to 7 as students count.)
- S: 4, 8, 12, 16, 20, 24, 28.
- T: What is 7 × 4?
- S: 28.
- T: Let's see how we can skip-count down to find the answer, too. (Show 10 fingers.)
       Start at 10 fours, 40. (Count down with your fingers as students say numbers.)
- S: 40, 36, 32, 28.

Lesson 2:   Decompose and recompose shapes to compare areas.

Continue with the following possible sequence: 9×4, 6×4, and 8×4.

T: (Distribute Multiply by 4 (6–10) Pattern Sheet.) Let's practice multiplying by 4. Be sure to work left to right across the page.

Directions for administration of a Multiply-By Pattern Sheet are as follows:

1. Distribute Pattern Sheet.
2. Allow a maximum of two minutes for students to complete as many problems as possible.
3. Direct students to work left to right across the page.
4. Encourage skip-counting strategies to solve unknown facts.

Application Problem (5 minutes)

Wilma and Freddie use pattern blocks to make shapes as shown. Freddie says his shape has a bigger area than Wilma's because it is longer than hers. Is he right? Explain your answer.

Wilma's Shape: 6 triangles
                     6 rhombuses
                     1 hexagon

No, Freddie is not right. They both use the same pattern blocks, but they arranged them differently. Since they used the same pattern blocks, their shapes have the same areas.

Freddie's Shape: 6 triangles
                     6 rhombuses
                     1 hexagon

Wilma's Shape

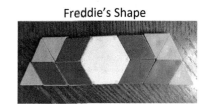

Freddie's Shape

Note: This problem reviews the Lesson 1 concept that, although shapes look different, they may have the same area.

Concept Development (34 minutes)

Materials: (S) Paper Strip 1: 1 in × 12 in, Paper Strip 2: 1 cm × 12 cm, scissors, ruler, Problem Set Page 1

Students begin with Paper Strip 1; it should be oriented with the long sides on the top and bottom.

T: Measure your strip. How tall is it?

S: 1 inch tall.

T: Start at the edge of your strip, and use your ruler to mark inches along the top. Do the same along the bottom. Use your ruler to connect the marks at the top to the matching marks at the bottom.

NOTES ON MULTIPLE MEANS OF ACTION AND EXPRESSION:

Make it easy for learners to mark inches and cut the strip using the following tips:

- Provide strips of thicker paper, such as cardstock.
- Provide strips of grid paper to facilitate drawing lines.
- If you offer paper strips with pre-drawn tick marks, guide discovery of *inches*. Darken lines for cutting.
- Offer left-handed and adaptive scissors, if needed.

T: How many units make up your strip?

S: 12 units.

T: What shape are they?

S: They're squares. Each of the 4 sides is 1 inch.

T: What is the area of the paper strip in square units?

S: 12 square units!

T: Because the sides of the squares each measure 1 inch, we call one of these squares a square inch. What is the area of your paper strip in square inches?

MP.6 S: 12 square inches.

T: Did the number of squares change?

S: No.

T: Talk to a partner. What changed about the way we talked about the area of the paper strip?

S: The units changed. → Before today, we called them square units, but now we can call them square inches because all 4 sides measure 1 inch. → We named this square unit. A square unit could have sides of any length. A square inch is always one inch on every side.

T: Cut your paper strip along the lines you drew. Now, rearrange all 12 squares into 2 equal rows. Remember, the squares have to touch but cannot overlap.

T: Draw your rectangle in the chart for Problem 1 on your Problem Set, next to where it says *Rectangle A.* (Model.) What is the area of the rectangle?

S: 12 square inches.

T: Record the area. You can record it by writing 12 square inches, or you can write 12 sq in.

T: Rearrange all 12 squares into 3 equal rows to make a new rectangle. Draw it in the chart for *Rectangle B* and record the area. At my signal, whisper the area of your rectangle to a partner. (Signal.)

S: 12 square inches.

T: Rearrange all 12 squares into 4 equal rows to make a new rectangle. Draw it in the chart for *Rectangle C* and record the area. At my signal, whisper the area of your rectangle to a partner. (Signal.)

S: 12 square inches.

T: How is it possible that these three different rectangles and our paper strip all have the same area?

S: We used the same squares for each one, so they all have the same area. → We rearranged 12 square inches each time. Just rearranging them doesn't change the area.

Repeat the process with Paper Strip 2 (1 cm × 12 cm).
Call attention to the change in units to centimeters. Discuss similarities and differences between the rectangular models. Students should notice that the same models can be built even though the units are different.

NOTES ON
MULTIPLE MEANS
OF ENGAGEMENT:

Students working above grade level may enjoy more autonomy as they explore and compare area. Offer the choice of a partner game in which Partner A constructs a shape, after which Partner B constructs a shape with a greater or lesser area. Encourage students to modify the game or invent another that compares area.

**EUREKA
MATH**™

Note: The square inch and square centimeter tiles will be used again in other Module 4 lessons. You may want to collect them or have students store them in a safe place.

Problem Set (10 minutes)

Students should do their personal best to complete the Problem Set within the allotted 10 minutes. Some problems do not specify a method for solving. This is an intentional reduction of scaffolding that invokes MP.5, Use Appropriate Tools Strategically. Students should solve these problems using the RDW approach used for Application Problems.

For some classes, it may be appropriate to modify the assignment by specifying which problems students should work on first. With this option, let the purposeful sequencing of the Problem Set guide the selections so that problems continue to be scaffolded. Balance word problems with other problem types to ensure a range of practice. Consider assigning incomplete problems for homework or at another time during the day.

Student Debrief (10 minutes)

Lesson Objective: Decompose and recompose shapes to compare areas.

The Student Debrief is intended to invite reflection and active processing of the total lesson experience.

Invite students to review their solutions for the Problem Set. They should check work by comparing answers with a partner before going over answers as a class. Look for misconceptions or misunderstandings that can be addressed in the Debrief. Guide students in a conversation to debrief the Problem Set and process the lesson.

Any combination of the questions below may be used to lead the discussion.

- Talk to a partner. What new units did we define today?

- Look at Problem 4. If Maggie uses square inches for Shape A and square centimeters for Shape B, which shape has a larger area? How do you know?

- Compare the shape you drew in Problem 5 to a partner's. Are they the same? Do they have the same area? Why or why not?
- We started our lesson by using an inch ruler to break apart a rectangle into square inches. Turn and talk to a partner. Why was it important to break apart the rectangle into square inches?

Exit Ticket (3 minutes)

After the Student Debrief, instruct students to complete the Exit Ticket. A review of their work will help with assessing students' understanding of the concepts that were presented in today's lesson and planning more effectively for future lessons. The questions may be read aloud to the students.

©2015 Great Minds. eureka-math.org
G3-M4-TE-B4-1.3.1-01.2016

Multiply.

4 x 1 = _____   4 x 2 = _____   4 x 3 = _____   4 x 4 = _____

4 x 5 = _____   4 x 6 = _____   4 x 7 = _____   4 x 8 = _____

4 x 9 = _____   4 x 10 = _____   4 x 6 = _____   4 x 7 = _____

4 x 6 = _____   4 x 8 = _____   4 x 6 = _____   4 x 9 = _____

4 x 6 = _____   4 x 10 = _____   4 x 6 = _____   4 x 7 = _____

4 x 6 = _____   4 x 7 = _____   4 x 8 = _____   4 x 7 = _____

4 x 9 = _____   4 x 7 = _____   4 x 10 = _____   4 x 7 = _____

4 x 8 = _____   4 x 6 = _____   4 x 8 = _____   4 x 7 = _____

4 x 8 = _____   4 x 9 = _____   4 x 8 = _____   4 x 10 = _____

4 x 8 = _____   4 x 9 = _____   4 x 6 = _____   4 x 9 = _____

4 x 7 = _____   4 x 9 = _____   4 x 8 = _____   4 x 9 = _____

4 x 10 = _____   4 x 9 = _____   4 x 10 = _____   4 x 6 = _____

4 x 10 = _____   4 x 7 = _____   4 x 10 = _____   4 x 8 = _____

4 x 10 = _____   4 x 9 = _____   4 x 10 = _____   4 x 6 = _____

4 x 8 = _____   4 x 10 = _____   4 x 7 = _____   4 x 9 = _____

multiply by 4 (6–10)

Lesson 2:   Decompose and recompose shapes to compare areas.   **25**

©2015 Great Minds. eureka math.org
G3-M4-TE-B4-1.3.1-01.2016 -

Name _____ Date _____

1. Use all of Paper Strip 1, which you cut into 12 square inches, to complete the chart below.

	Drawing	Area
Rectangle A		
Rectangle B		
Rectangle C		

2. Use all of Paper Strip 2, which you cut into 12 square centimeters, to complete the chart below.

	Drawing	Area
Rectangle A		
Rectangle B		
Rectangle C		

Lesson 2: Decompose and recompose shapes to compare areas.

EUREKA MATH™

3. Compare the areas of the rectangles you made with Paper Strip 1 and Paper Strip 2. What changed? Why did it change?

4. Maggie uses square units to create these two rectangles. Do the two rectangles have the same area? How do you know?

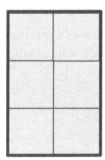

Shape A

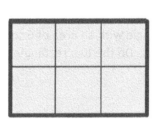

Shape B

5. Count to find the area of the rectangle below. Then, draw a different rectangle that has the same area.

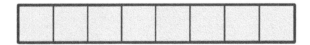

Name _____   Date _____

1. Each [ ] is a square unit. Find the area of the rectangle below. Then, **draw a different rectangle** with the same number of square units.

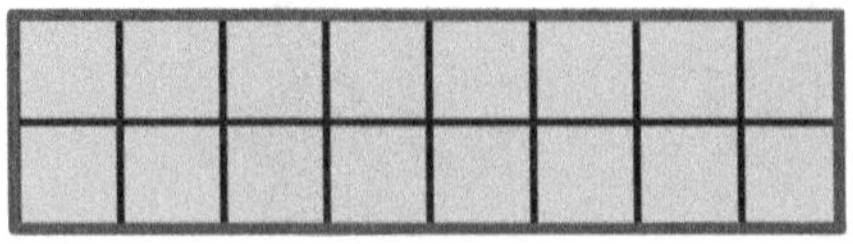

2. Zach creates a rectangle with an area of 6 square inches. Luke makes a rectangle with an area of 6 square centimeters. Do the two rectangles have the same area? Why or why not?

EUREKA
MATH™

Name _____   Date _____

1. Each ☐ is a square unit.   Count to find the area of each rectangle.   Then, circle all the rectangles with an area of 12 square units.

a.

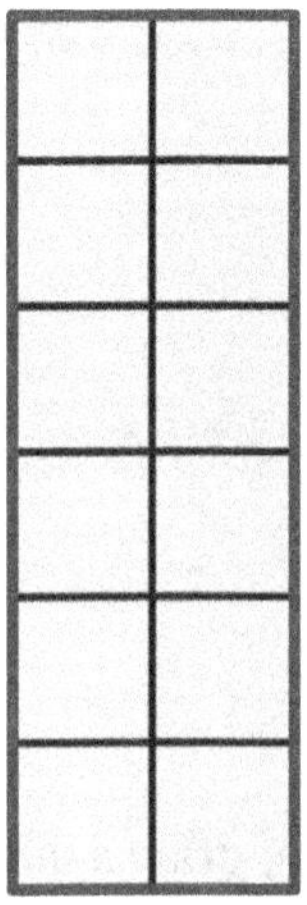

Area = _____ square units

b.

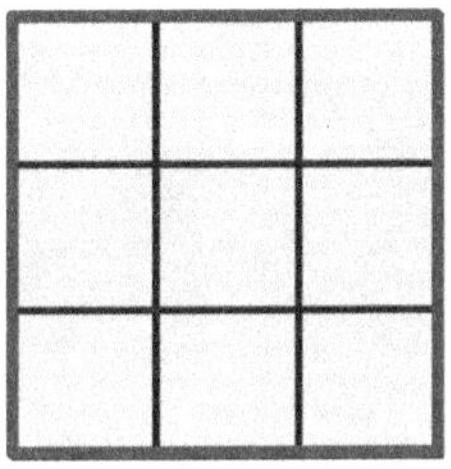

Area = _____ square units

c.

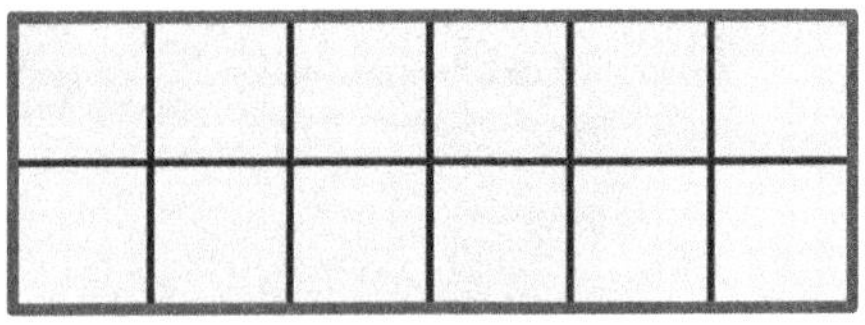

Area = _____ square units

d.

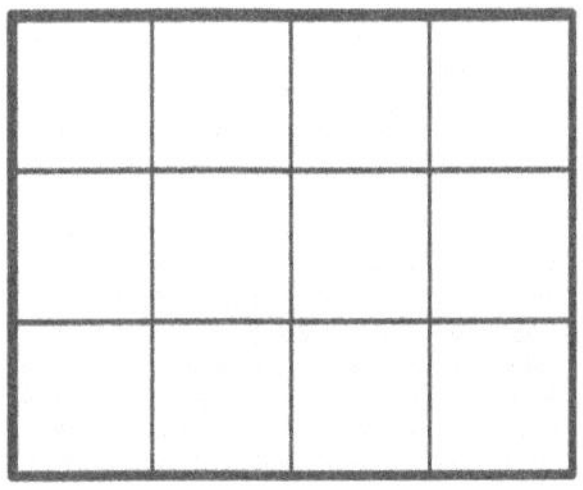

Area = _____ square units

e.

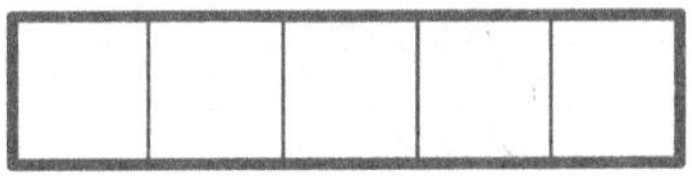

Area = _____ square units

f.

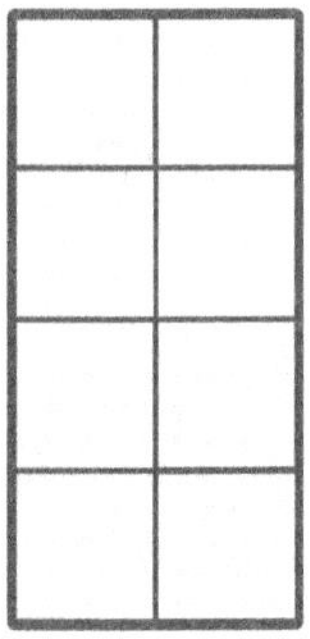

Area = _____ square units

2.   Colin uses square units to create these rectangles. Do they have the same area? Explain.

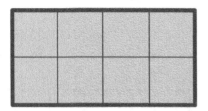   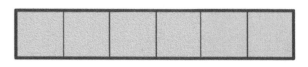

3.   Each 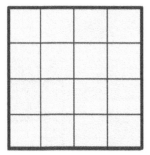 is a square unit. Count to find the area of the rectangle below. Then, draw a different rectangle that has the same area.

©2015 Great Minds. eureka-math.org
G3-M4-TE-B4-1.3.1-01.2016

EUREKA
MATH™

Lesson 3

Objective: Model tiling with centimeter and inch unit squares as a strategy to measure area.

Suggested Lesson Structure

■ Fluency Practice   (13 minutes)
▨ Application Problem   (5 minutes)
▢ Concept Development   (32 minutes)
▨ Student Debrief   (10 minutes)

   Total Time   **(60 minutes)**

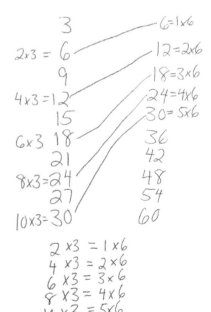

Fluency Practice (13 minutes)

- Find the Common Products **3.OA.7**   (7 minutes)
- Count the Square Units **3.MD.6**   (6 minutes)

Find the Common Products (7 minutes)

Materials: (S) Blank paper

Note: This fluency activity reviews multiplication patterns from Module 3.

   T: Fold your paper in half vertically. Unfold your paper. On the left half, count by threes to 30 down the side of your paper. On the right half, count by sixes to 60 down the side of your paper. Draw a line to match the products that appear in both columns.

   S: (Match 6, 12, 18, 24, and 30.)

   T: (Write ___ × 3 = 6, ___ × 3 = 12, ___ × 3 = 18, ___ × 3 = 24, and ___ × 3 = 30 next to each matched product on the left half of the paper.) Write the equations next to their products like I did, completing the unknown factors.

   S: (Write equations and complete unknowns.)

   T: (Write 6 = ___ × 6, 12 = ___ × 6, 18 = ___ × 6, 24 = ___ × 6, and 30 = ___ × 6 next to each matched product on the left half of the paper.) Write the equations next to their products like I did, completing the unknown factors.

   S: (Write equations and complete unknowns.)

T: (Write 2 × 3 = ___ × 6.) Say the equation, completing the unknown factor.

S: 2 × 3 = 1 × 6.

T: (Write 2 × 3 = 1 × 6.) Write the remaining equal facts as equations.

S: (Write 4 × 3 = 2 × 6, 6 × 3 = 3 × 6, 8 × 3 = 4 × 6, and 10 × 3 = 5 × 6.)

T: What is the pattern in your equations?

S: Each multiple of 6 is also a multiple of 3.

Count the Square Units (6 minutes)

Note: This fluency activity reviews finding total area using square units.

T: (Project a 1 × 5 tiled array similar to Figure 1 on the right.) What's the area of the rectangle? (Pause.)

S: 5 square units.

Continue with Figures 2–5.

Figures for Count the Square Units

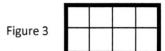
Figure 1

Figure 2

Figure 3

Figure 4

Figure 5

Application Problem (5 minutes)

Jace uses paper squares to create a rectangle. Clary cuts all of Jace's squares in half to create triangles. She uses all the triangles to make a rectangle. There are 16 triangles in Clary's rectangle. How many squares were in Jace's shape?

The following are possible student solutions:

- Dividing

  16 ÷ 2 = 8   There were 8 squares in Jace's shape.

- Drawing a picture

  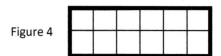   16 ÷ 2 = 8

- Skip-counting by twos

  2, 4, 6, 8, 10, 12, 14, 16   8 twos   16 ÷ 2 = 8

Note: This problem reviews multiplying or dividing by units of 2 from Module 1, depending on how students solve. Invite students to share their strategies for solving.

32   Lesson 3:   Model tiling with centimeter and inch unit squares as a strategy to measure area.

©2015 Great Minds. eureka-math.org
G3-M4-TE-B4-1.3.1-01.2016

EUREKA
MATH

Concept Development (32 minutes)

Centimeter Grid   Inch Grid

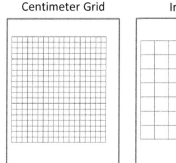

Materials: (S) Square centimeter and square inch tiles (from Lesson 2), centimeter grid (Template 1) and inch grid (Template 2), ruler, personal white board

Pass out 10 square centimeter tiles to each student.

T: Arrange all of your square tiles in 2 equal rows to create a rectangle. Make sure the tiles are touching and do not overlap. (Allow students time to create a rectangle.) What is the area of your rectangle?

S: 10 square units.

T: Is there another way you could arrange all of your tiles to make a rectangle?

S: We could make 5 rows of 2. → Or, 1 row of 10.

T: Make 1 row of 10. (Allow students time to make a new rectangle.) What is the area of your rectangle now?

S: It is still 10 square units!

T: Use your ruler to measure all four sides of a tile in centimeters. (Wait for students to measure.) Can we define these units more precisely?

S: Yes, they're square centimeters. → Yes, all four sides measure 1 centimeter, so they are square centimeters.

T: What is the area of your rectangle in square centimeters?

S: 10 square centimeters.

T: (Pass out the centimeter grid.) Slip the grid paper into your personal white board. Each side of the squares in the grid measures 1 centimeter. How is this grid paper like the tiles we used?

S: They are both square centimeters.

T: Shade the grid paper to represent the rectangle you made with tiles.

S: (Shade grid paper.)

T: Remove a tile from your rectangle, making sure your tiles all still touch to form a rectangle. (Pause.) What is the area of the rectangle now?

S: 9 square centimeters!

T: How can you change the rectangle on the grid paper to have the same area as your new tile rectangle?

S: Erase one of the squares.

NOTES ON MULTIPLE MEANS OF ACTION AND EXPRESSION:

Offer an alternative to drawing, shading, and erasing rectangles using a marker. Some students may find it easier to represent and shade rectangles using an interactive white board or personal computer.

NOTES ON MULTIPLE MEANS OF ACTION AND EXPRESSION:

Support English language learners as they compose their written response to Problem 3. Discussing their reasoning with a partner before writing may be advantageous. Encourage students to use *area* and *square units* in their response. Request that students clarify, if necessary, and guide the elaboration of their ideas.

EUREKA MATH™

Lesson 3:   Model tiling with centimeter and inch unit squares as a strategy to measure area.

©2015 Great Minds. eureka math.org
G3-M4-TE-B4-1.3.1-01.2016

33

T: Go ahead and do that. (Students erase a square.) What is the area of the shaded rectangle?

S: 9 square centimeters.

Repeat this process using the inch grid and inch tiles. If time allows, students can shade a shape for a partner, who then finds the area of the shape. Then, they can erase squares to create shapes with smaller areas. As students are ready, they can begin drawing shapes using squares rather than just erasing them.

Problem Set (10 minutes)

Inch and centimeter grid paper are required for some of these problems. Students should do their personal best to complete the Problem Set within the allotted 10 minutes. For some classes, it may be appropriate to modify the assignment by specifying which problems they work on first. Some problems do not specify a method for solving. Students should solve these problems using the RDW approach used for Application Problems.

Student Debrief (10 minutes)

Lesson Objective: Model tiling with centimeter and inch unit squares as a strategy to measure area.

The Student Debrief is intended to invite reflection and active processing of the total lesson experience.

Invite students to review their solutions for the Problem Set. They should check work by comparing answers with a partner before going over answers as a class. Look for misconceptions or misunderstandings that can be addressed in the Debrief. Guide students in a conversation to debrief the Problem Set and process the lesson.

Any combination of the questions below may be used to lead the discussion.

- How are the rectangles in Problems 1(b) and 1(c) the same? How are they different?

- How are the rectangles in Problems 1(a) and 2(a) the same? How are they different?

- Which rectangle in Problem 2 has the largest area? How do you know?

- Compare the rectangles you made in Problem 4 with a partner's rectangles. How are they the same? How are they different?

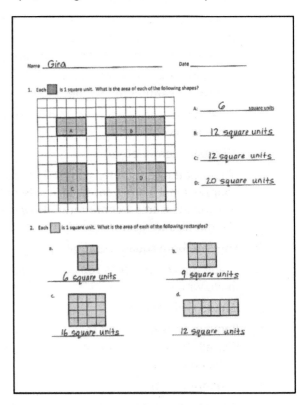

Lesson 3:   Model tiling with centimeter and inch unit squares as a strategy to measure area.

©2015 Great Minds. eureka-math.org
G3-M4-TE-B4-1.3.1-01.2016

EUREKA MATH

Exit Ticket (3 minutes)

After the Student Debrief, instruct students to complete the Exit Ticket. A review of their work will help with assessing students' understanding of the concepts that were presented in today's lesson and planning more effectively for future lessons. The questions may be read aloud to the students.

3. a. How would the rectangles in Problem 1 be different if they were composed of square inches?

The shapes in Problem 1 would be bigger if they were made of square inches. The number of squares would stay the same, but the size of the squares would change.

b. Select one rectangle from Problem 1 and recreate it on square inch and square-centimeter grid paper.

(See attached example)

4. Use a separate piece of square-centimeter grid paper. Draw four different rectangles that each has an area of 8 square centimeters.

(See attached example)

Examples of Problem 3(b) and Problem 4

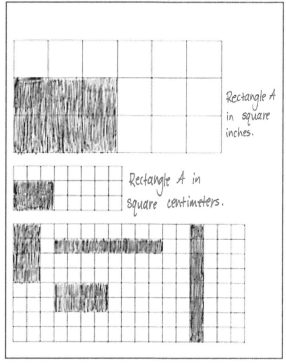

Rectangle A in square inches.

Rectangle A in square centimeters.

Lesson 3:   Model tiling with centimeter and inch unit squares as a strategy to measure area.

©2015 Great Minds. eureka math.org
G3-M4-TE-B4-1.3.1-01.2016

35

Name _____ Date _____

1. Each ☐ is 1 square unit. What is the area of each of the following rectangles?

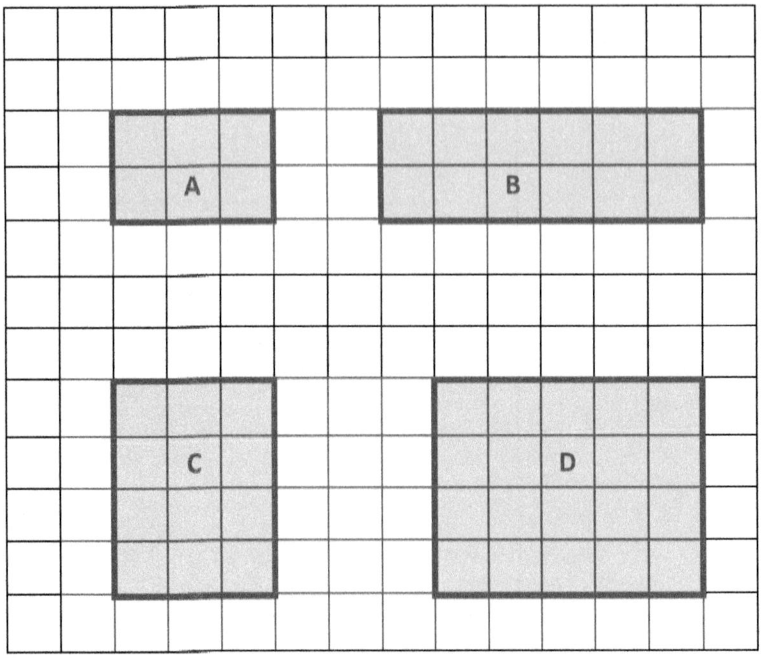

A: _____ square units

B: _____

C: _____

D: _____

2. Each ☐ is 1 square unit. What is the area of each of the following rectangles?

a.

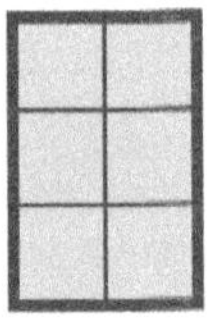

_____

b.

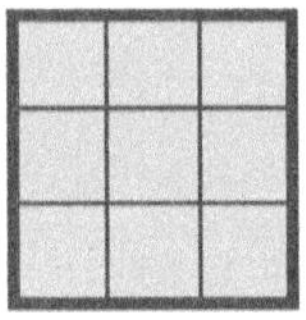

_____

c.

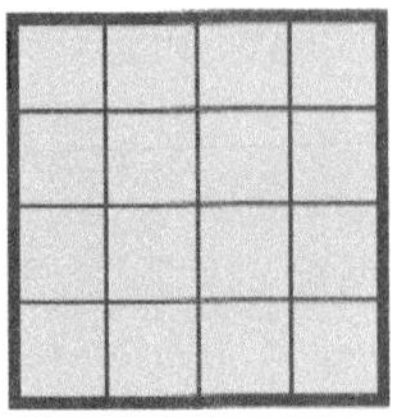

_____

d.

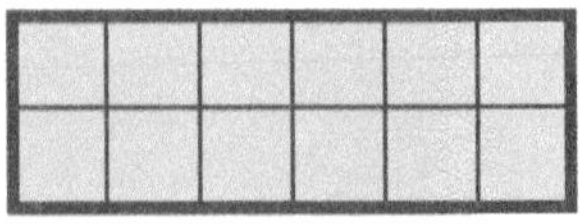

_____

Lesson 3:   Model tiling with centimeter and inch unit squares as a strategy to measure area.

©2015 Great Minds. eureka-math.org
G3-M4-TE-B4-1.3.1-01.2016

EUREKA MATH™

3. a. How would the rectangles in Problem 1 be different if they were composed of square inches?

    b. Select one rectangle from Problem 1 and recreate it on square inch and square centimeter grid paper.

4. Use a separate piece of square centimeter grid paper. Draw four different rectangles that each has an area of 8 square centimeters.

 Lesson 3:   Model tiling with centimeter and inch unit squares as a strategy to measure area.

©2015 Great Minds. eureka math.org
G3-M4-TE-B4-1.3.1-01.2016 -

37

Name _____   Date _____

1. Each ▢ is 1 square unit. Write the area of Rectangle A. Then, draw a different rectangle with the same area in the space provided.

Area = _____

2. Each ▢ is 1 square unit. Does this rectangle have the same area as Rectangle A? Explain.

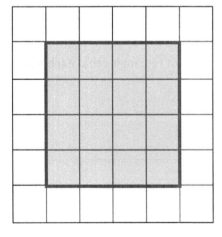

Lesson 3:   Model tiling with centimeter and inch unit squares as a strategy to measure area.

©2015 Great Minds. eureka-math.org
G3-M4-TE-B4-1.3.1-01.2016

EUREKA MATH

Name _____ Date _____

1. Each ☐ is 1 square unit. What is the area of each of the following rectangles?

A: _____ square units

B: _____

C: _____

D: _____

2. Each ☐ is 1 square unit. What is the area of each of the following **rectangles**?

a.

b.

c.

d.

Lesson 3: Model tiling with centimeter and inch unit squares as a strategy to measure area.

©2015 Great Minds. eureka math.org
G3-M4-TE-B4-1.3.1-01.2016

39

3. Each ☐ is 1 square unit. Write the area of each rectangle. Then, draw a different rectangle with the same area in the space provided.

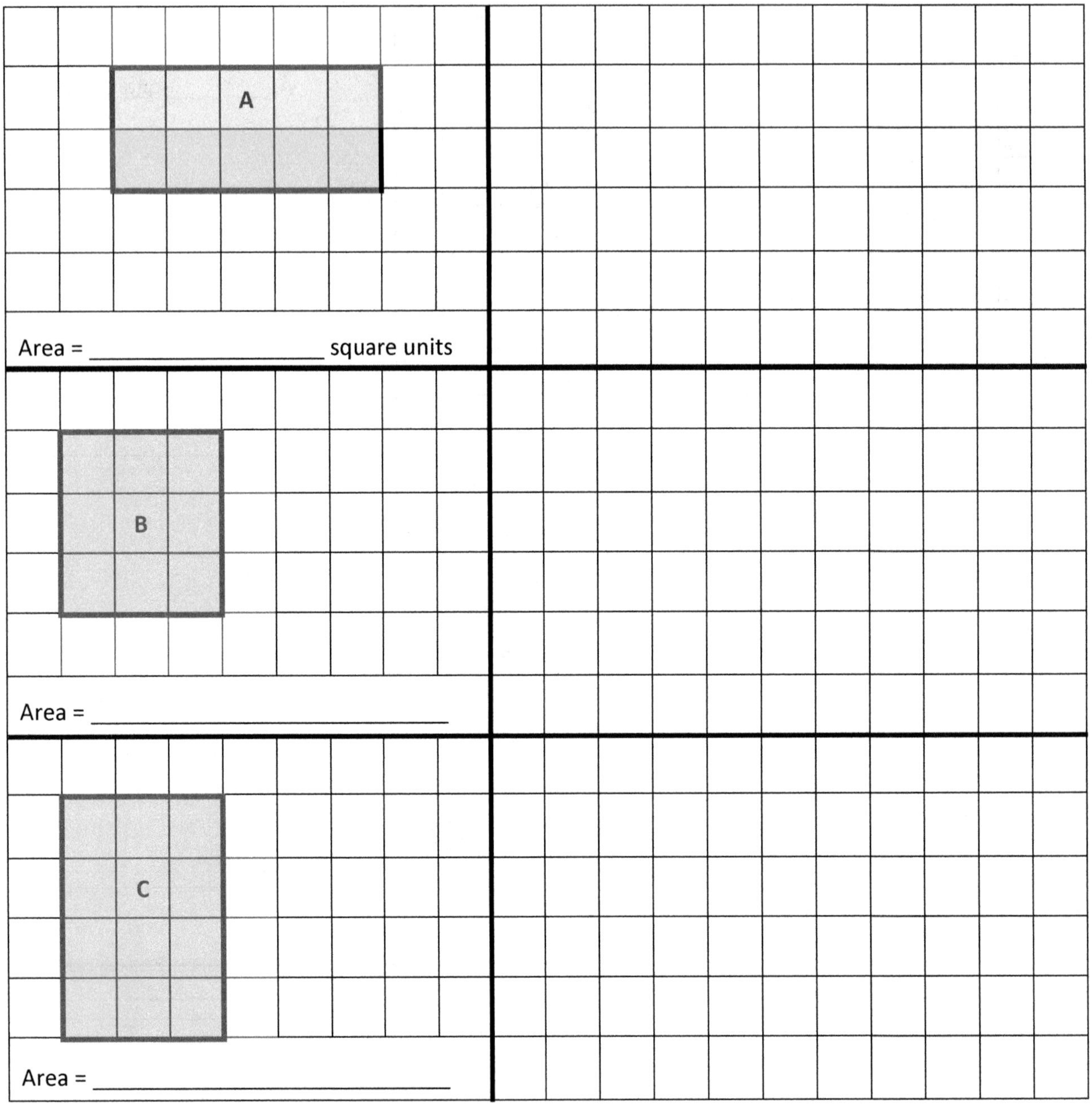

A

Area = _____ square units

B

Area = _____

C

Area = _____

Lesson 3: Model tiling with centimeter and inch unit squares as a strategy to measure area.

©2015 Great Minds. eureka-math.org
G3-M4-TE-B4-1.3.1-01.2016

EUREKA MATH™

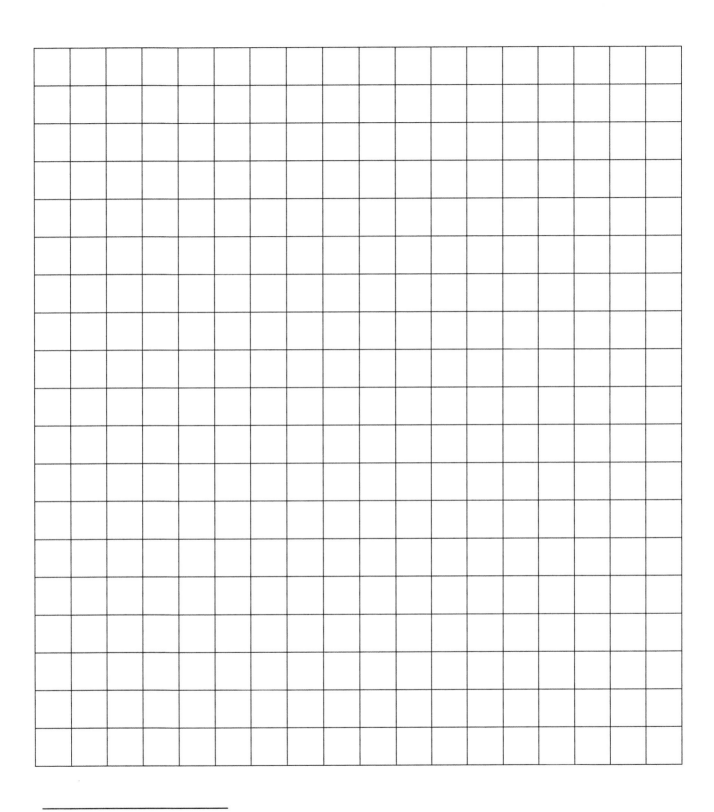

centimeter grid

Lesson 3:   Model tiling with centimeter and inch unit squares as a strategy to
measure area.

©2015 Great Minds. eureka math.org
G3-M4-TE-B4-1.3.1-01.2016 -

41

inch grid

Lesson 3:   Model tiling with centimeter and inch unit squares as a strategy to
measure area.

Lesson 4

Objective: Relate side lengths with the number of tiles on a side.

Suggested Lesson Structure

■ Fluency Practice   (12 minutes)
▨ Application Problem   (5 minutes)
▢ Concept Development   (33 minutes)
■ Student Debrief   (10 minutes)

   Total Time   **(60 minutes)**

Fluency Practice (12 minutes)

- Group Counting **3.OA.1**   (3 minutes)
- Products in an Array **3.OA.3**   (3 minutes)
- Count the Square Units **3.MD.6**   (6 minutes)

Group Counting (3 minutes)

Note: Group counting reviews interpreting multiplication as repeated addition.

Instruct students to count forward and backward, occasionally changing the direction of the count.

- Sixes to 60
- Sevens to 70
- Eights to 80
- Nines to 90

Products in an Array (3 minutes)

Materials: (S) Personal white board

Note: This fluency activity anticipates relating multiplication with area in Topic B.

T: (Project an array with 5 rows of 3 stars.) How many rows of stars do you see?

S: 5 rows.

T: How many stars are in each row?

S: 3 stars.

T: On your personal white board, write two different multiplication sentences that can be used to find the total number of stars.

S: (Write $5 \times 3 = 15$ and $3 \times 5 = 15$.)

Continue with the following possible sequence: 4×6, 7×3, 8×5, and 9×7.

Count the Square Units (6 minutes)

Materials: (T) 12 square tiles

Note: This fluency activity reviews comparing the area of different shapes.

- T: (Project an 8 × 1 tiled array.) How many square units are in the rectangle?
- S: 8 square units.
- T: (Write 8 *square units* next to the rectangle. Project a 4 × 2 tiled array.) How many square units are in the rectangle?
- S: 8 square units.
- T: (Write 8 *square units* next to the rectangle. Project a 2 × 4 tiled array.) How many square units are in the rectangle?
- S: 8 square units.
- T: (Write 8 *square units* next to the rectangle. Project a 1 × 8 tiled array.) How many square units are in the rectangle?
- S: 8 square units.
- T: (Write 8 *square units* next to the rectangle.) Do the four rectangles look the same?
- S: No.
- T: What do the rectangles have in common?
- S: They are each composed of 8 square units.

Continue with the following possible sequence: 12 × 1, 1 × 12, 6 × 2, 3 × 4, 2 × 6, and 4 × 3.

Application Problem (5 minutes)

Mara uses 15 square-centimeter tiles to make a rectangle. Ashton uses 9 square-centimeter tiles to make a rectangle.

- a.   Draw what Mara and Ashton's rectangles might look like.
- b.   Whose rectangle has a bigger area? How do you know?

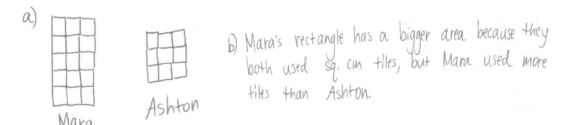

a)

Mara

Ashton

b) Mara's rectangle has a bigger area because they both used sq. cm tiles, but Mara used more tiles than Ashton.

Note: This problem reviews Lesson 2, particularly tiling with square units. Invite students to share and compare their drawings for Mara and Ashton's rectangles.

44

Lesson 4:   Relate side lengths with the number of tiles on a side.

©2015 Great Minds. eureka-math.org
G3-M4-TE-B4-1.3.1-01.2016

EUREKA
MATH

Concept Development (33 minutes)

Materials: (S) 15 square inch and square centimeter tiles, ruler, personal white board

Note: The *ones* cubes included in sets of Dienes blocks (base ten blocks) may also be used as square centimeter tiles.

Pass out 15 square-inch tiles to each student.

T: These tiles are square…?

S: Inches!

T: Use the tiles to make a 3 by 5 array. (Allow students time to make an array.) Push the tiles together to form a rectangle with no gaps or overlaps. What is the area of your rectangle?

S: 15 square inches.

T: I see your squares are nicely arranged to form a rectangle. What about these? (Project Rectangles A and B shown to the right.) I used 15 square-inch tiles to make both of these rectangles. Talk to a partner. Is the area of both rectangles 15 square inches?

Rectangle A

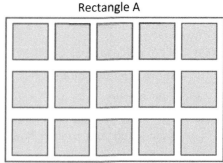

S: Yes. The number of tiles is the same. → No. A's area is bigger than 15 square inches because there are gaps between the tiles. B's area is smaller because some of the tiles are on top of each other.

Rectangle B

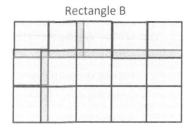

T: Why is it important to avoid gaps or overlaps when we measure area?

S: If there are gaps or overlaps, the amount of space the rectangle takes up changes. → The square unit would be wrong because some area is taken away if there are overlaps, or some is added if there are gaps.

T: Use your ruler to measure across the top of your rectangle in inches. What is the length of this side?

S: 5 inches.

T: How many tiles are on this side?

S: 5 tiles.

T: Use your ruler to measure the shorter side of the rectangle in inches. What is the length of this side?

S: 3 inches.

 MP.8   T: How many tiles are on this side?

S: 3 tiles!

T: What is the relationship between the number of tiles on a side and the side length of the rectangle?

S: They are the same.

NOTES ON
MULTIPLE MEANS
OF REPRESENTATION:

Scaffold student contrast of *length* and *area*. Consider placing a long string along the side of the rectangle, or have students trace the side with a finger to better illustrate *length*. In contrast, have students shade in the *area* before writing 15 square inches.

 EUREKA
MATH™

T: What do you notice about the lengths of the opposite sides of the rectangles?

S: They are equal.

T: Trace the rectangle on your personal white board. Then, remove the tiles and label the side lengths. Now, write the area inside the rectangle. What are the units for the side lengths?

S: Inches.

T: What are the units for the area?

S: Square inches.

T: Talk to a partner. Why are the units different for side lengths and area?

S: The unit for side lengths is inches because we used a ruler to measure the length of the side in inches. For area, the unit is square inches because we counted the number of square inch tiles that we used to make the rectangle.

T: Inches are used to measure lengths, such as the side lengths, and square inches are used to measure the amount of flat space a figure takes up, which is the area.

Direct students to exchange square inch tiles for square centimeter tiles.

T: These tiles are square…?

S: Centimeters.

T: Use them to make a rectangle with side lengths of 5 centimeters and 4 centimeters. (Write 5 cm and 4 cm.) Tell your partner how many tiles you will count to make each side.

S: I will make one side with 5 tiles and the other with 4 tiles. → Actually, we will count 5 tiles each for two sides of the rectangle and 4 tiles each for the other two sides. Opposite sides are the same, remember?

T: Make your rectangle on top of your board. Label the side lengths.

S: (Make rectangle and label side lengths 5 cm and 4 cm.)

T: How many fives did you make? Why?

S: 4 fives because the other side length is 4.

T: What is the total of 4 fives?

S: 20.

T: Skip-count your fives to find the total area of the rectangle. (Pause.) What is the total area?

S: 20 square centimeters.

T: What is the relationship between the side lengths and area?

S: If you multiply 5 times 4, then you get 20.

If time allows, repeat the process using a rectangle with side lengths of 3 centimeters and 6 centimeters. As students are ready, tell them the area, and let them build a rectangle and name the side lengths.

Problem Set (10 minutes)

Students should do their personal best to complete the Problem Set within the allotted 10 minutes. For some classes, it may be appropriate to modify the assignment by specifying which problems they work on first. Some problems do not specify a method for solving. Students should solve these problems using the RDW approach used for Application Problems.

Student Debrief (10 minutes)

Lesson Objective: Relate side lengths with the number of tiles on a side.

The Student Debrief is intended to invite reflection and active processing of the total lesson experience.

Invite students to review their solutions for the Problem Set. They should check work by comparing answers with a partner before going over answers as a class. Look for misconceptions or misunderstandings that can be addressed in the Debrief. Guide students in a conversation to debrief the Problem Set and process the lesson.

Any combination of the questions below may be used to lead the discussion.

- Tell a partner how you could use square centimeter tiles to check your work in Problem 1.

- Compare the areas of the rectangles in Problems 1 and 2. Which rectangle has a larger area? How do you know?

- What are the side lengths of the shape in Problem 3? Are all of the sides the same? How do you know? What shape is this?

- What is the area of the rectangle in Problem 4? Explain how you found the area to a partner.

- How many centimeter tiles fit in the rectangle in Problem 5? Is that the area of the rectangle in square centimeters? Why or why not?

- In Problem 6, if the side length of A is 4 units, would 3 units make sense for the side length of B? Why or why not? What *would* make sense?

Lesson 4:   Relate side lengths with the number of tiles on a side.

47

©2015 Great Minds. eureka math.org
G3-M4-TE-B4-1.3.1-01.2016 -

Exit Ticket (3 minutes)

After the Student Debrief, instruct students to complete the Exit Ticket. A review of their work will help with assessing students' understanding of the concepts that were presented in today's lesson and planning more effectively for future lessons. The questions may be read aloud to the students.

Name _____ Date _____

1. Use a ruler to measure the side lengths of the rectangle in centimeters. Mark each centimeter with a point and connect the points to show the square units. Then, count the squares you drew to find the total area.

Total area: _____

2. Use a ruler to measure the side lengths of the rectangle in inches. Mark each inch with a point and connect the points to show the square units. Then, count the squares you drew to find the total area.

Total area: _____

3. Mariana uses square centimeter tiles to find the side lengths of the rectangle below. Label each side length. Then, count the tiles to find the total area.

Total area: _____

Lesson 4:   Relate side lengths with the number of tiles on a side.

©2015 Great Minds. eureka math.org
G3-M4-TE-B4-1.3.1-01.2016

49

4. Each 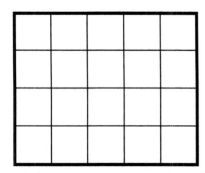 is 1 square centimeter. Saffron says that the side length of the rectangle below is 4 centimeters. Kevin says the side length is 5 centimeters. Who is correct? Explain how you know.

5. Use both square centimeter and square inch tiles to find the area of the rectangle below. Which works best? Explain why.

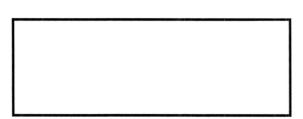

6. How does knowing side lengths A and B help you find side lengths C and D on the rectangle below?

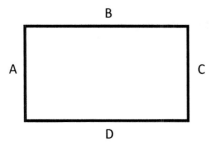

Lesson 4:   Relate side lengths with the number of tiles on a side.

©2015 Great Minds. eureka-math.org
G3-M4-TE-B4-1.3.1-01.2016

EUREKA
MATH

Name _____ Date _____

Label the side lengths of each rectangle. Then, match the rectangle to its total area.

a.

12 square centimeters

b.

5 square inches

c.

6 square centimeters

Lesson 4: Relate side lengths with the number of tiles on a side.

©2015 Great Minds. eureka math.org
G3-M4-TE-B4-1.3.1-01.2016 -

51

Name _____   Date _____

1. Ella placed square centimeter tiles on the rectangle below, and then labeled the side lengths. What is the area of her rectangle?

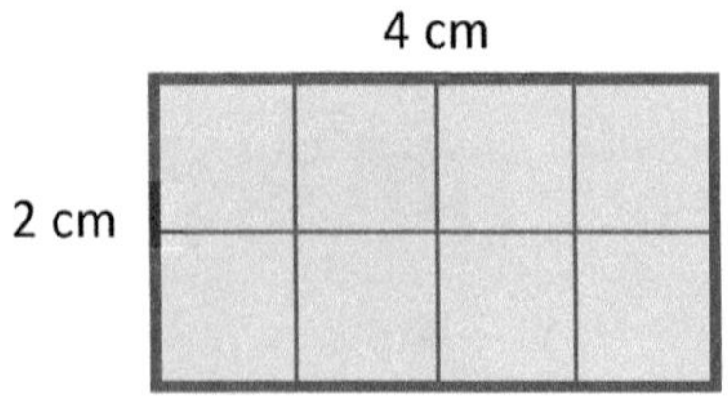

4 cm

2 cm

Total area: _____

2. Kyle uses square centimeter tiles to find the side lengths of the rectangle below. Label each side length. Then, count the tiles to find the total area.

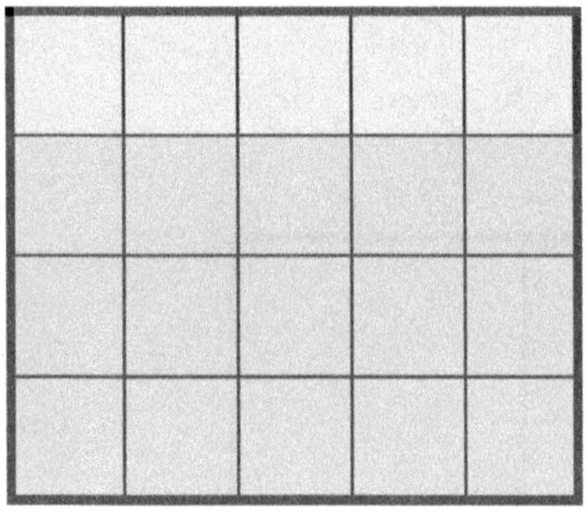

Total area: _____

3. Maura uses square inch tiles to find the side lengths of the rectangle below. Label each side length. Then, find the total area.

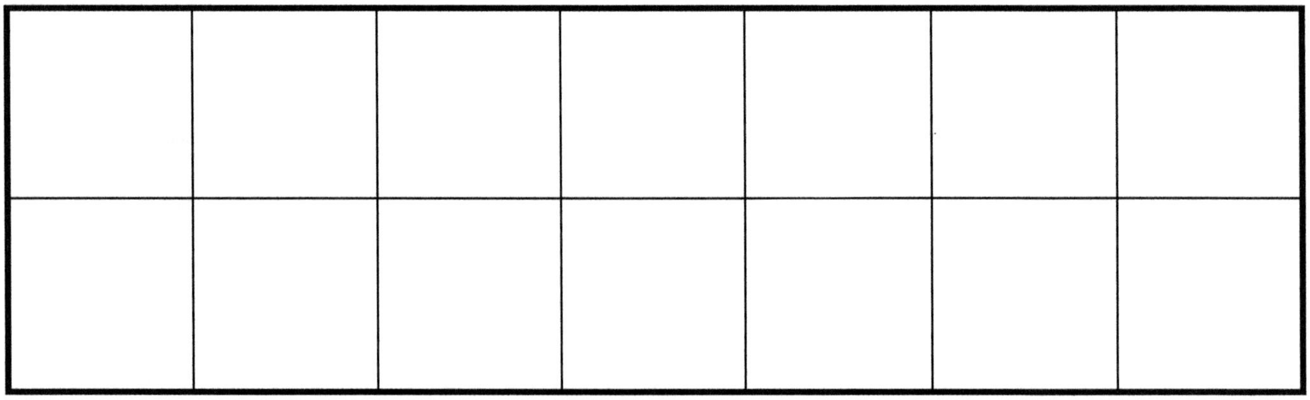

Total area: _____

Lesson 4:   Relate side lengths with the number of tiles on a side.

EUREKA
MATH

4. Each square unit below is 1 square inch. Claire says that the side length of the rectangle below is 3 inches. Tyler says the side length is 5 inches. Who is correct? Explain how you know.

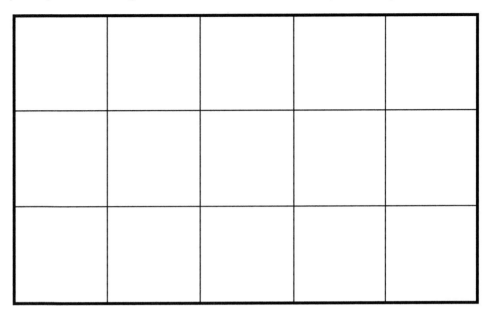

5. Label the unknown side lengths for the rectangle below, and then find the area. Explain how you used the lengths provided to find the unknown lengths and area.

4 inches

2 inches

_____

_____

Total area: _____

Lesson 4:   Relate side lengths with the number of tiles on a side.

©2015 Great Minds. eureka math.org
G3-M4-TE-B4-1.3.1-01.2016 -

53

Mathematics Curriculum

3
GRADE

Topic B

Concepts of Area Measurement

3.MD.5, 3.MD.6, 3.MD.7abd

Focus Standards:	3.MD.5	Recognize area as an attribute of plane figures and understand concepts of area measurement.
		a. A square with side length 1 unit, called "a unit square," is said to have "one square unit" of area, and can be used to measure area.
		b. A plane figure which can be covered without gaps or overlaps by *n* unit squares is said to have an area of *n* square units.
	3.MD.6	Measure areas by counting unit squares (square cm, square m, square in, square ft, and improvised units).
	3.MD.7	Relate area to the operations of multiplication and addition.
		a. Find the area of a rectangle with whole-number side lengths by tiling it, and show that the area is the same as would be found by multiplying the side lengths.
		b. Multiply side lengths to find areas of rectangles with whole-number side lengths in the context of solving real world and mathematical problems, and represent whole-number products as rectangular areas in mathematical reasoning.
		d. Recognize area as additive. Find areas of rectilinear figures by decomposing them into non-overlapping rectangles and adding the areas of the non-overlapping parts, applying this technique to solve real world problems.
Instructional Days:	4	
Coherence -Links from:	G2–M2	Addition and Subtraction of Length Units
	G3–M1	Properties of Multiplication and Division and Solving Problems with Units of 2–5 and 10
	G3–M3	Multiplication and Division with Units of 0, 1, 6–9, and Multiples of 10
-Links to:	G4–M3	Multi-Digit Multiplication and Division
	G4–M7	Exploring Multiplication

In previous lessons, students tiled given rectangles. In Lesson 5, students build rectangles using unit square tiles to make arrays when provided with specific criteria. For example, students may be told that there are 24 tiles inside the rectangle and one side of the rectangle is covered with 4 tiles. Students may start by building one column of the array to represent a length of 4 units and then duplicate that process until they reach 24 total tiles, skip-counting by fours. Finally, they physically push together the rows of tiles to make the array. When they count the number of fours, the process connects to unknown factor problems (in this case, the unknown factor of 6) from previous modules and builds toward students' discovery of the area formula.

Now experienced with drawing rectangular arrays within an area model, students find the area of an incomplete array in Lesson 6. They visualize and predict what the finished array looks like and then complete it by joining opposite end points with a straight edge. They determine the total area using skip-counting. The incomplete array model bridges to the area model, where no array is given.

In Lesson 7, students receive information about the side lengths of an area model (shown to the right). Based on this information, they use a straight edge to draw a grid of equal-sized squares within the area model and then skip-count to find the total number of squares. Units move beyond square centimeters and inches to include square feet and square meters.

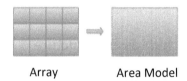

Array   Area Model

In Lesson 8, students recognize that side lengths play an important part in determining the area of a rectangle. They understand that multiplying the number of square units in a row by the number of rows produces the same result as skip-counting the squares within the array. Given the area and one side length, students realize that they can use multiplication with an unknown factor or division to find the unknown side length.

A Teaching Sequence Toward Mastery of Concepts of Area Measurement

Objective 1: Form rectangles by tiling with unit squares to make arrays.
(Lesson 5)

Objective 2: Draw rows and columns to determine the area of a rectangle given an incomplete array.
(Lesson 6)

Objective 3: Interpret area models to form rectangular arrays.
(Lesson 7)

Objective 4: Find the area of a rectangle through multiplication of the side lengths.
(Lesson 8)

©2015 Great Minds. eureka math.org
G3-M4-TE-B4-1.3.1-01.2016 -

Lesson 5

Objective: Form rectangles by tiling with unit squares to make arrays.

Suggested Lesson Structure

■ Fluency Practice   (14 minutes)
■ Application Problem   (6 minutes)
□ Concept Development (30 minutes)
■ Student Debrief   (10 minutes)
    Total Time   **(60 minutes)**

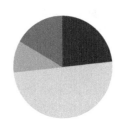

Fluency Practice (14 minutes)

- Group Counting **3.OA.1**   (3 minutes)
- Products in an Array **3.OA.3**   (3 minutes)
- Find the Common Products **3.OA.7**   (8 minutes)

Group Counting (3 minutes)

Note: Group counting reviews interpreting multiplication as repeated addition.

Instruct students to count forward and backward, occasionally changing the direction of the count.

- Threes to 30
- Sixes to 60
- Sevens to 70
- Nines to 90

Products in an Array (3 minutes)

Materials: (S) Personal white board

Note: This fluency activity anticipates relating multiplication with area in Topic B's lessons.

    T: (Project an array with 4 rows of 3 stars.) How many rows of stars do you see?
    S: 4 rows.
    T: How many stars are in each row?
    S: 3 stars.
    T: On your personal white board, write two multiplication sentences that can be used to find the total number of stars.
    S: (Write $4 \times 3 = 12$ and $3 \times 4 = 12$.)

Continue with the following possible sequence: 3×6, 7×5, 8×4, and 9×6.

©2015 Great Minds. eureka-math.org
G3-M4-TE-B4-1.3.1-01.2016

Find the Common Products (8 minutes)

Materials: (S) Blank paper

Note: This fluency activity reviews multiplication patterns from Module 3.

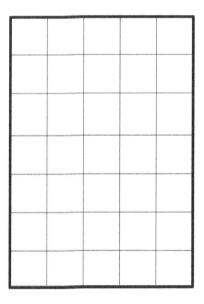

T: (List the multiples of 4 and 8 in two different columns.) Draw a line to match the products that appear in both columns.

S: (Match 8, 16, 24, 32, and 40.)

T: (Write 2 × 4 = 8, etc., next to each matched product on the left half of the paper.) Write the rest of the equations like I did.

S: (Write equations.)

T: (Write 8 = 1 × 8, etc., next to each matched product on the right half of the paper.) Write the rest of the equations like I did.

S: (Write equations.)

T: (Write 2 × 4 = ___ × 8.) Say the true equation.

S: 2 × 4 = 1 × 8.

T: (Write 2 × 4 = 1 × 8.) Write the remaining equal facts as equations.

S: (Write 4 × 4 = 2 × 8, 6 × 4 = 3 × 8, 8 × 4 = 4 × 8, and 10 × 4 = 5 × 8.)

T: Discuss the patterns in your equations.

S: Each multiple of 8 is also a multiple of 4.

Application Problem (6 minutes)

Candice uses square centimeter tiles to find the side lengths of a rectangle as shown on the right. She says the side lengths are 5 centimeters and 7 centimeters. Her partner, Luis, uses a ruler to check Candice's work and says that the side lengths are 5 centimeters and 6 centimeters. Who is right? How do you know?

Candice is right because she used square centimeter tiles to find the side lengths and when I counted the tiles there were 5 on one side 7 on the other side. That means that the side lengths are 5 cm and 7 cm.

Note: This problem reviews Lesson 4, particularly the relationship between the number of tiles and side length. Invite students to discuss what Luis might have done wrong.

Lesson 5:   Form rectangles by tiling with unit squares to make arrays.

57

Concept Development (30 minutes)

Materials: (S) 15 square inch tiles, personal white board, straight edge, blank paper

Concrete: Understand the relationship between side lengths and area.

Draw or project the rectangle and side length shown on the right.

2 in

- T: Use square inch tiles to show this rectangle as an array. What information do we know?
- S: There are 2 rows. → A side length is 2 inches.
- T: At your table, place tiles to make the known side.
- S: (Make 1 column of 2 tiles.)
- T: (Write below the diagram: *Area = 12 square inches.*) How many total tiles will we use to make our rectangle?
- S: 12 tiles.
- T: How many twos are in 12?
- S: 6 twos.
- T: Use your tiles to make 6 sets of twos, and then skip-count to check your work.
- S: (Make 6 groups of 2 tiles and skip-count.) 2, 4, 6, 8, 10, 12.
- T: Push your twos together to make a rectangle. (Allow students time to complete. Add a question mark to the diagram as shown on the right.) What is the unknown side length?
- S: Six. → Six tiles. → Six inches.
- T: (Replace the question mark with *6 in* on the diagram.) Tell your partner about the relationship between the side lengths and the area. Write an equation to show your thinking. Be sure to include the units.
- S: 2 inches × 6 inches = 12 square inches, so the area is the product of the side lengths. (Write *2 inches × 6 inches = 12 square inches.*)

> **NOTES ON MULTIPLE MEANS OF REPRESENTATION:**
>
> Simplify and clarify your script for English language learners and others. Rephrase, "What information do we know?" to "How many rows of inch squares? How do you know?"
>
> Place the opaque 2 by 6 rectangle pictured above over a square grid. As an alternate way to check work, lift the rectangle to show the 12 squares covered.

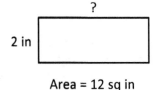

?

2 in

Area = 12 sq in

Repeat the process using a rectangle with a known side length of 5 inches and an area of 15 square inches. Ask students to write an unknown factor problem, 5 × ___ = 15, and then use the tiles to solve.

58   **Lesson 5:**   Form rectangles by tiling with unit squares to make arrays.

©2015 Great Minds. eureka-math.org
G3-M4-TE-B4-1.3.1-01.2016

EUREKA
MATH

Concrete/Pictorial: Form rectangles and determine area or side lengths by drawing to make arrays.

T: Lay tiles on your personal white board to make a side 3 inches tall. Trace the outline of all 3 tiles. Then, draw horizontal lines to show where they connect.

S: (Draw a 3-inch tall side.)

T: Label the side length.

S: (Label *3 in* as shown.)

T: Use your tiles to make another side 7 inches long.

S: (Add tiles horizontally, using the top tile as one of the 7.)

T: Trace the outline of the tiles. Draw vertical lines to show where they connect. Label the side length.

S: (Draw image shown on the right and label *7 in*.)

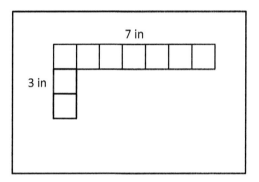

T: How many sets of threes will be in this rectangle?

S: 7 threes.

T: Talk to your partner. Which strategy might you use to find the total area of the rectangle?

S: We can draw in the rest of the squares and count them all. → Or, just skip-count 7 threes. → It would be easier to just multiply 7 inches × 3 inches and get 21 square inches.

T: Many students suggested multiplying the side lengths to find the area. Let's check this strategy by drawing in the rest of the squares. Use your straight edge to draw the rest of the tiles in the rectangle, and then skip-count to find the total area.

S: (Draw the rest of the tiles, and then skip-count.) 3, 6, 9, 12, 15, 18, 21.

T: Does 7 inches × 3 inches = 21 square inches accurately give the area of the rectangle?

S: Yes!

T: Clear your personal white board, and use your tiles to make a side length of 6 inches. Trace the outline of all 6 tiles. Then, draw horizontal lines to show where they connect.

S: (Draw image shown to the right.)

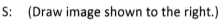

T: Label the side length.

S: (Label *6 in* as shown on the right.)

T: Write 6 × __ = 24 on your personal white board. Talk to a partner about how you can use this equation to help find the other side length.

S: From the equation, I know that the area is 24, so I can add rows of 6 tiles until I have 24 tiles. Then, I can count the rows to find the side length. → I can skip-count by 6 to get to 24, and then I know the other side length will be equal to the number of times I skip-count. → I know 6 × 4 = 24, so I know that the other side length is 4.

T: Choose a strategy to find the other side length, and then fill in the blank in the equation. (Allow time for students to work.) What is the other side length?

S: 4 inches!

Lesson 5:   Form rectangles by tiling with unit squares to make arrays.

©2015 Great Minds. eureka-math.org
G3-M4-TE-B4-1.3.1-01.2016 -

59

Problem Set (10 minutes)

Students should do their personal best to complete the Problem Set within the allotted 10 minutes. For some classes, it may be appropriate to modify the assignment by specifying which problems they work on first. Some problems do not specify a method for solving. Students should solve these problems using the RDW approach used for Application Problems.

Student Debrief (10 minutes)

Lesson Objective: Form rectangles by tiling with unit squares to make arrays.

The Student Debrief is intended to invite reflection and active processing of the total lesson experience.

Invite students to review their solutions for the Problem Set. They should check their work by comparing answers with a partner before going over answers as a class. Look for misconceptions or misunderstandings that can be addressed in the Debrief. Guide students in a conversation to debrief the Problem Set and process the lesson.

Any combination of the questions below may be used to lead the discussion.

- Compare Problems 1(b) and 1(e) and Problems 1(a) and 1(c). How does each pair show commutativity?

- How many more threes does the array in Problem 1(d) have compared to the array in Problem 1(a)? How might the side lengths help you know that, even without seeing the tiled array?

MP.8

- Compare Problems 1(c) and 1(f). How are the areas related? (The area of 1(f) is half the area of 1(c).) How might you have figured that out just by knowing the side lengths of each array?

- In Problem 2, what strategy did you use to find the unknown side length? Is there another way you could have figured it out?

- Students may have different solutions for Problem 3. Invite them to share and compare their work.

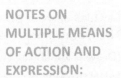

NOTES ON MULTIPLE MEANS OF ACTION AND EXPRESSION:

Some learners may benefit from alternatives to drawing tiles inside rectangles on the Problem Set. Consider the following:

- Magnify the Problem Set to ease small motor tasks.

- Provide virtual or concrete manipulatives.

- Allow students to draw their own rectangles, perhaps with larger tiles, perhaps with smaller areas.

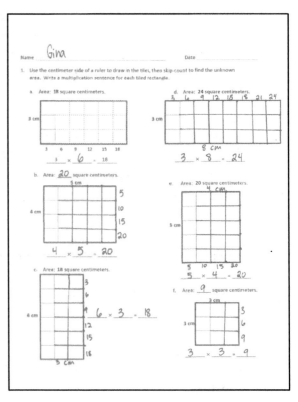

EUREKA MATH™

Exit Ticket (3 minutes)

After the Student Debrief, instruct students to complete the Exit Ticket. A review of their work will help with assessing students' understanding of the concepts that were presented in today's lesson and planning more effectively for future lessons. The questions may be read aloud to the students.

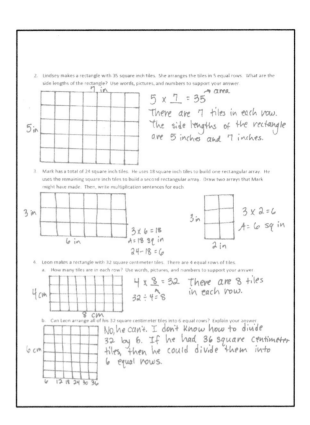

EUREKA MATH

Lesson 5:   Form rectangles by tiling with unit squares to make arrays.

61

©2015 Great Minds. eureka math.org
G3-M4-TE-B4-1.3.1-01.2016

Name _____   Date _____

1. Use the centimeter side of a ruler to draw in the tiles, and then skip-count to find the unknown area.
    Write a multiplication sentence for each tiled rectangle.

a. Area: **18** square centimeters.

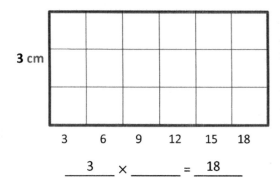

3 cm

| 3 | 6 | 9 | 12 | 15 | 18 |

___3___ × _____ = __18__

b. Area: _____ square centimeters.

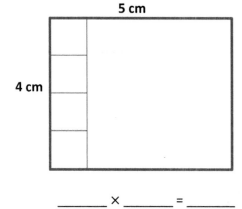

5 cm

4 cm

_____ × _____ = _____

c. Area: **18** square centimeters.

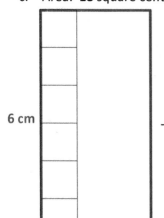

6 cm

_____ × _____ = _____

d. Area: **24** square centimeters.

3 cm

_____ × _____ = _____

e. Area: **20** square centimeters.

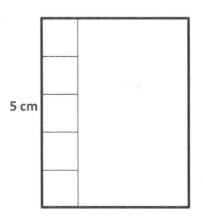

5 cm

_____ × _____ = _____

f. Area: _____ square centimeters.

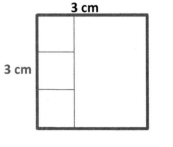

3 cm

3 cm

_____ × _____ = _____

Lesson 5:   Form rectangles by tiling with unit squares to make arrays.

EUREKA
MATH™

©2015 Great Minds. eureka-math.org
G3-M4-TE-B4-1.3.1-01.2016

2. Lindsey makes a rectangle with 35 square inch tiles. She arranges the tiles in 5 equal rows. What are the side lengths of the rectangle? Use words, pictures, and numbers to support your answer.

3. Mark has a total of 24 square inch tiles. He uses 18 square inch tiles to build one rectangular array. He uses the remaining square inch tiles to build a second rectangular array. Draw two arrays that Mark might have made. Then, write multiplication sentences for each.

4. Leon makes a rectangle with 32 square centimeter tiles. There are 4 equal rows of tiles.

   a. How many tiles are in each row? Use words, pictures, and numbers to support your answer.

   b. Can Leon arrange all of his 32 square centimeter tiles into 6 equal rows? Explain your answer.

EUREKA MATH

Lesson 5:   Form rectangles by tiling with unit squares to make arrays.

63

©2015 Great Minds. eureka math.org
G3-M4-TE-B4-1.3.1-01.2016 -

Name _____ Date _____

Darren has a total of 28 square centimeter tiles. He arranges them into 7 equal rows. Draw Darren's rectangle. Label the side lengths, and write a multiplication sentence to find the total area.

Lesson 5:   Form rectangles by tiling with unit squares to make arrays.

©2015 Great Minds. eureka-math.org
G3-M4-TE-B4-1.3.1-01.2016

EUREKA
MATH™

Name _____ Date _____

1. Use the centimeter side of a ruler to draw in the tiles, and then skip-count to find the unknown area. Write a multiplication sentence for each tiled rectangle.

a. Area: **24** square centimeters.

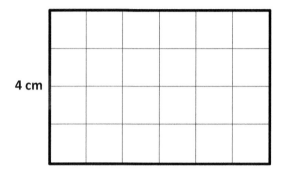

____4____ × _____ = ___24___

b. Area: **24** square centimeters.

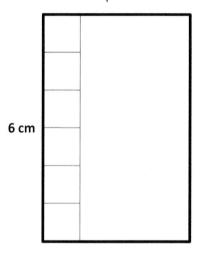

_____ × _____ = _____

c. Area: **15** square centimeters.

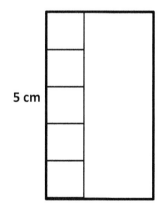

_____ × _____ = _____

d. Area: **15** square centimeters.

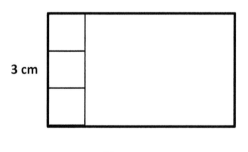

_____ × _____ = _____

EUREKA MATH™

Lesson 5:   Form rectangles by tiling with unit squares to make arrays.

65

©2015 Great Minds. eureka math.org
G3-M4-TE-B4-1.3.1-01.2016   -

2. Ally makes a rectangle with 45 square inch tiles. She arranges the tiles in 5 equal rows. How many square inch tiles are in each row? Use words, pictures, and numbers to support your answer.

3. Leon makes a rectangle with 36 square centimeter tiles. There are 4 equal rows of tiles.

    a. How many tiles are in each row? Use words, pictures, and numbers to support your answer.

    b. Can Leon arrange all of his 36 square centimeter tiles into 6 equal rows? Use words, pictures, and numbers to support your answer.

    c. Do the rectangles in Parts (a) and (b) have the same total area? Explain how you know.

66   Lesson 5:   Form rectangles by tiling with unit squares to make arrays.

EUREKA
MATH

Lesson 6

Objective: Draw rows and columns to determine the area of a rectangle given an incomplete array.

Suggested Lesson Structure

■ Fluency Practice   (12 minutes)
▨ Application Problem   (8 minutes)
▨ Concept Development   (30 minutes)
■ Student Debrief   (10 minutes)

   Total Time   **(60 minutes)**

Fluency Practice (12 minutes)

- Group Counting **3.OA.1**   (4 minutes)
- Write the Multiplication Fact **3.MD.7**   (4 minutes)
- Products in an Array **3.OA.3**   (4 minutes)

Group Counting (4 minutes)

Note: Group counting reviews interpreting multiplication as repeated addition.

Instruct students to count forward and backward, occasionally changing the direction of the count.

- Sixes to 60
- Sevens to 70
- Eights to 80
- Nines to 90

Write the Multiplication Fact (4 minutes)

Materials: (S) Personal white board

Note: This fluency activity reviews relating multiplication with area from Lesson 5.

T: (Project a 5 by 3 square unit tiled rectangle. Write ___ × ___ = 15.) There are 15 tiles altogether.
    How many rows are there?
S: 5 rows.
T: (Write 5 × ___ = 15.) On your personal white board, fill in the blank to make the equation true.
S: (Write 5 × 3 = 15.)

T: (Project a 3 by 4 square unit tiled rectangle. Write ___ × ___ = 12.) There are 12 tiles altogether. How many columns are there?

S: 4 columns.

T: (Write ___ × 4 = 12.) On your personal white board, fill in the blank to make the equation true.

S: (Write 3 × 4 = 12.)

Continue with the following possible sequence, asking students to first name either the number of rows or the number of columns: 4 × 6, 6 × 7, 5 × 8, and 7 × 8.

Products in an Array (4 minutes)

Materials: (S) Personal white board

Note: This fluency activity supports the relationship between multiplication and area.

T: (Project an array with 2 rows of 6 stars.) How many rows of stars do you see?

S: 2 rows.

T: How many stars are in each row?

S: 6 stars.

T: On your personal white board, write two multiplication sentences that can be used to find the total number of stars.

S: (Write 2 × 6 = 12 and 6 × 2 = 12.)

Continue with the following possible sequence: 3 × 7, 6 × 5, 8 × 6, and 4 × 9.

Application Problem (8 minutes)

Huma has 4 bags of square inch tiles with 6 tiles in each bag. She uses them to measure the area of a rectangle on her homework. After covering the rectangle, Huma has 4 tiles left. What is the area of the rectangle?

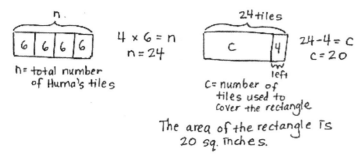

NOTES ON
MULTIPLE MEANS
OF ENGAGEMENT:

Adjust the numbers in the Application Problem to challenge students working above grade level.

Note: This problem reviews multi-step word problems in the context of using square tiles to measure area.

Lesson 6:   Draw rows and columns to determine the area of a rectangle given an incomplete array.

©2015 Great Minds. eureka-math.org
G3-M4-TE-B4-1.3.1-01.2016

EUREKA
MATH™

Concept Development (30 minutes)

Materials: (S) Personal white board, straight edge, Problem Set, array 1 (Template 1), array 2 (Template 2)

NOTES ON MULTIPLE MEANS FOR ACTION AND EXPRESSION:

Part 1: Estimate to draw the missing square units inside an array.

Students have Templates 1 and 2 in their personal white boards, and are looking at array 1.

Scaffold the following sequence further by beginning with a basic 2 by 2 rectangle in which 2 tiles are missing. Graduate to a 2 by 3 rectangle in which tiles or lines are missing. Continue step by step until students are ready for rectangles with larger areas. Also, consider adding color to alternating tiles to assist with counting or distinguishing tiles from rectangles or blank space.

- T: How can an array of square units help you find the area of a rectangle?
- S: You can count the total number of square units inside the rectangle. → You can skip-count the rows to find the total.
- T: (Project or display the image on the right.) What do you notice about the array inside of this rectangle?
- S: Some of the square units are missing.
- T: What do you notice about the top row?
- S: It has 4 square units and a rectangle.

Array 1

- T: Look at the second row. Can you use those square units to help you know how many square units make the top row?
- S: The second row has 1 more square unit than the top row. You can just follow the line it makes to divide the rectangle into 2 square units.
- T: Use your straight edge to draw that line now.
- S: (Draw as shown on the right.)
- T: Talk to your partner. Use the top row to figure out how many square units will fit in each of the rows below. How do you know?

MP.2

- S: Each row should have 6 square units because rows in an array are equal.

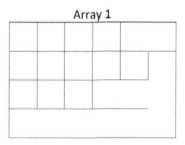
Array 1: Top Row Complete

- T: Use the lines that are already there as guides, and with your straight edge, draw lines to complete the array.
- S: (Draw.)
- T: How many rows of 6 are in this array?
- S: 4 rows of 6.
- T: What equation can be used to find the area of the rectangle?
- S: $4 \times 6 = 24$.

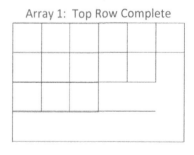
Array 1: Fully Drawn

Part 2: Draw rows and columns to determine the area.

T: (Project the rectangle shown on the right.) Look at array 2. Can we estimate to draw **unit squares** inside the rectangle?

S: Yes.

T: It might take us longer because fewer units are given. A quicker way to find the area is to figure out the number of rows and the number of columns. Let's start by finding the number of rows in our array. How can we find the number of rows?

S: The first column shows you how many rows there are.

T: With your finger, show your partner what you will draw to find the number of rows. Then, draw.

S: (Show and draw.)

T: How can we find the number of columns?

S: The first row shows you how many columns there are.

T: Use your straight edge to complete the first row. Label the side lengths of the rectangle, including units.

S: (Draw and label side lengths 5 units and 6 units.)

T: What number sentence can be used to find the area?

S: 5 × 6 = 30.

Array 2

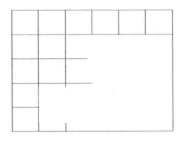

Array 2: 1 Row and
1 Column Drawn

Problem Set (10 minutes)

Students should do their personal best to complete the Problem Set within the allotted 10 minutes. For some classes, it may be appropriate to modify the assignment by specifying which problems they work on first. Some problems do not specify a method for solving. Students should solve these problems using the RDW approach used for Application Problems.

Student Debrief (10 minutes)

Lesson Objective: Draw rows and columns to determine the area of a rectangle given an incomplete array.

The Student Debrief is intended to invite reflection and active processing of the total lesson experience.

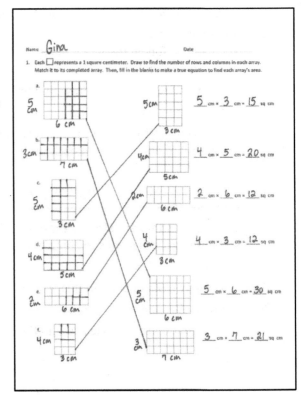

Lesson 6: Draw rows and columns to determine the area of a rectangle given an incomplete array.

©2015 Great Minds. eureka-math.org
G3-M4-TE-B4-1.3.1-01.2016

EUREKA
MATH

Invite students to review their solutions for the Problem Set. They should check work by comparing answers with a partner before going over answers as a class. Look for misconceptions or misunderstandings that can be addressed in the Debrief. Guide students in a conversation to debrief the Problem Set and process the lesson.

Any combination of the questions below may be used to lead the discussion.

- How did you know where to draw the columns and rows in Problem 1?
- To find area, why is it not necessary to draw all of the **unit squares** in an incomplete array?
- What mistake did Sheena make in Problem 2?
- Is it necessary to have the rug to solve Problem 3? Why or why not?
- In Problem 3, how many tiles does the rug touch?
- There are multiple ways to find a solution to Problem 4. Invite students to share how they found the answer.

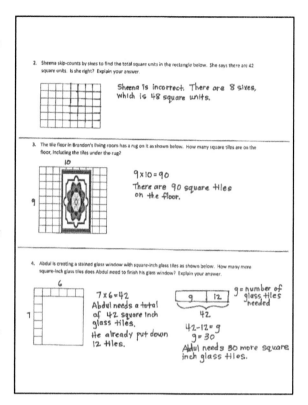

Exit Ticket (3 minutes)

After the Student Debrief, instruct students to complete the Exit Ticket. A review of their work will help with assessing students' understanding of the concepts that were presented in today's lesson and planning more effectively for future lessons. The questions may be read aloud to the students.

Lesson 6:   Draw rows and columns to determine the area of a rectangle given an
             incomplete array.

©2015 Great Minds. eureka math.org
G3-M4-TE-B4-1.3.1-01.2016 -

71

Name _____ Date _____

1. Each ☐ represents 1 square centimeter. Draw to find the number of rows and columns in each array. Match it to its completed array. Then, fill in the blanks to make a true equation to find each array's area.

a. 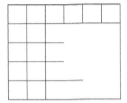      _____ cm × _____ cm = _____ sq cm

b.       _____ cm × _____ cm = _____ sq cm

c. 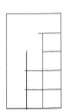   _____ cm × _____ cm = _____ sq cm

d. 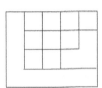   _____ cm × _____ cm = _____ sq cm

e. 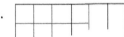   _____ cm × _____ cm = _____ sq cm

f.      _____ cm × _____ cm = _____ sq cm

Lesson 6: Draw rows and columns to determine the area of a rectangle given an incomplete array.

©2015 Great Minds. eureka-math.org
G3-M4-TE-B4-1.3.1-01.2016

EUREKA MATH

2. Sheena skip-counts by sixes to find the total square units in the rectangle below. She says there are 42 square units. Is she right? Explain your answer.

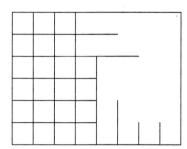

3. The tile floor in Brandon's living room has a rug on it as shown below. How many square tiles are on the floor, including the tiles under the rug?

4. Abdul is creating a stained glass window with square inch glass tiles as shown below. How many more square inch glass tiles does Abdul need to finish his glass window? Explain your answer.

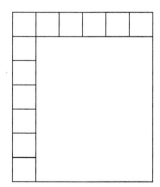

EUREKA MATH™

Lesson 6:   Draw rows and columns to determine the area of a rectangle given an incomplete array.

©2015 Great Minds. eureka math.org
G3-M4-TE-B4-1.3.1-01.2016 -

73

Name _____   Date _____

The tiled floor in Cayden's dining room has a rug on it as shown below. How many square tiles are on the floor, including the tiles under the rug?

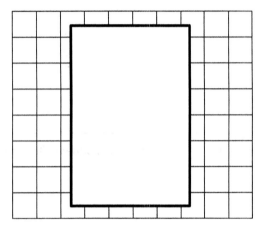

Lesson 6:   Draw rows and columns to determine the area of a rectangle given an
                          incomplete array.
                          ©2015 Great Minds. eureka-math.org
                          G3-M4-TE-B4-1.3.1-01.2016

EUREKA
MATH™

Name _____   Date _____

1. Each ☐ represents 1 square centimeter. Draw to find the number of rows and columns in each array.
    Match it to its completed array. Then, fill in the blanks to make a true equation to find each array's area.

a.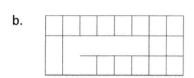

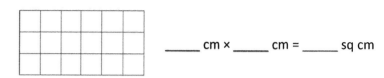   _____ cm × _____ cm = _____ sq cm

b.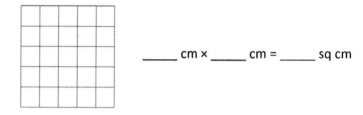

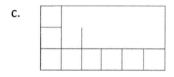   _____ cm × _____ cm = _____ sq cm

c.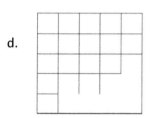

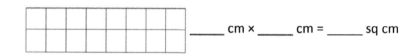   _____ cm × _____ cm = _____ sq cm

   _____ cm × _____ cm = _____ sq cm

d.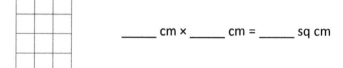

e. 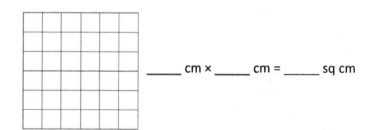   _____ cm × _____ cm = _____ sq cm

f.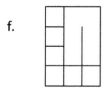

_____ cm × _____ cm = _____ sq cm

EUREKA
MATH™

Lesson 6:   Draw rows and columns to determine the area of a rectangle given an
             incomplete array.

©2015 Great Minds. eureka math.org
G3-M4-TE-B4-1.3.1-01.2016   -

75

2. Minh skip-counts by sixes to find the total square units in the rectangle below. She says there are 36 square units. Is she correct? Explain your answer.

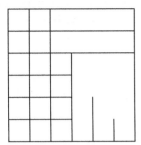

3. The tub in Paige's bathroom covers the tile floor as shown below. How many square tiles are on the floor, including the tiles under the tub?

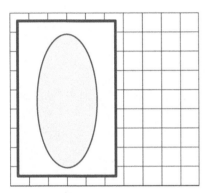

4. Frank sees a book on top of his chessboard. How many squares are covered by the book? Explain your answer.

Lesson 6:   Draw rows and columns to determine the area of a rectangle given an incomplete array.

©2015 Great Minds. eureka-math.org
G3-M4-TE-B4-1.3.1-01.2016

EUREKA
MATH™

array 1

Lesson 6: Draw rows and columns to determine the area of a rectangle given an incomplete array.

©2015 Great Minds. eureka math.org
G3-M4-TE-B4-1.3.1-01.2016 -

77

array 2

Lesson 6: Draw rows and columns to determine the area of a rectangle given an incomplete array.

EUREKA
MATH™

Lesson 7

Objective: Interpret area models to form rectangular arrays.

Suggested Lesson Structure

■ Fluency Practice   (12 minutes)
▨ Application Problem   (8 minutes)
▨ Concept Development   (30 minutes)
▨ Student Debrief   (10 minutes)
   Total Time   **(60 minutes)**

Fluency Practice (12 minutes)

- Group Counting **3.OA.1**   (4 minutes)
- Draw Rectangles **3.MD.5**   (4 minutes)
- Draw Rectangular Arrays **3.MD.5**   (4 minutes)

Group Counting (4 minutes)

Note: Group counting reviews interpreting multiplication as repeated addition.

Instruct students to count forward and backward, occasionally changing the direction of the count.

- Sixes to 60
- Sevens to 70
- Eights to 80
- Nines to 90

Draw Rectangles (4 minutes)

Materials: (S) Grid paper

Note: This fluency activity reviews drawing a rectangle from a known area. Show student work that is correct but looks different (e.g., a 6 × 2 unit rectangle juxtaposed with a 4 × 3 unit rectangle).

T: Draw a rectangle that has an area of 6 square units.
S: (Draw a 6-square unit rectangle.)

Continue with the following possible sequence: 10 square units, 12 square units, 16 square units, 24 square units, and 35 square units.

Draw Rectangular Arrays (4 minutes)

Materials: (S) Grid paper

Note: This fluency activity reviews finding area using side lengths.

    T: Draw a 4 × 2 rectangular array using the squares on your grid paper.
    T: How many square units are in your array?
    S: 8 square units.

Continue with the following possible sequence of rectangular arrays: 6 × 2, 4 × 3, 6 × 3, 9 × 2, 6 × 4, and 3 × 8.

Application Problem (8 minutes)

Lori wants to replace the square tiles on her wall. The square tiles are sold in boxes of 8 square tiles. Lori buys 6 boxes of tiles. Does she have enough to replace all of the tiles, including the tiles under the painting? Explain your answer.

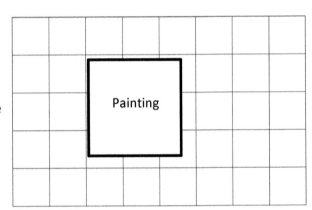

8 × 6 = 48

She bought 48 square tiles.

5 × 8 = 40

The area of the wall is 40 square tiles.

Yes, Lori will have enough tiles because she only needs 40 tiles, but she bought 48 tiles.

Note: This problem reviews multi-step word problems in the context of using square tiles to measure area. It also reviews finding the area of an incomplete array from Lesson 6.

Concept Development (30 minutes)

Materials: (T) Meter stick, 12-inch ruler, pad of square sticky notes (S) 1 set of square centimeter and square inch tiles per pair (from Lesson 2), personal white board, ruler, area model (Template)

Part 1: Explore the relationship between units and area.

    T: One partner will use square inches, and the other will use square centimeters. Work together to decide how to arrange your tiles to make the same shape rectangle. Then, create that rectangle with your pieces.
    S: (Decide on a rectangle and represent it using square inches and square centimeters.)

T: You and your partner each made the same shape rectangle. Is the area also the same?

S: We used the same number of pieces, but my pieces are smaller than yours. They are square centimeters, and look, my shape takes up less space on the table. → The area of the shape with square inches is bigger because inches are bigger than centimeters.

T: Turn your personal white board horizontally and write the area of your rectangle.

S: (Write either 12 square inches or 12 square centimeters.)

T: (Draw 1 square meter on the board.) This is 1 square meter. Suppose you used 12 square meter tiles to make your rectangle instead. Would this rectangle have a bigger area or a smaller area compared to your original rectangle?

MP.6   S: It would be much larger!

T: (Draw 1 square foot on the board.) How would your rectangle compare if you made it from 12 square feet?

S: It would be bigger than 12 square inches or centimeters but smaller than 12 square meters.

T: (Hold up a pad of square sticky notes.) How about if you had used 12 sticky notes?

S: Still bigger than 12 square inches or centimeters but smaller than 12 square feet or meters.

T: Why is it important to label the unit when you are talking about area?

S: Because how much area there is changes if the unit is small or big. → If you do not know the unit, then you do not really know what the area means. → It is just like with length. Twelve of a shorter unit is shorter than 12 of a longer unit.

Part 2: Relate area to multiplication to draw rectangular arrays.

T: Let us draw a rectangular array with an area of 18 square centimeters. How might we find the side lengths?

S: We could use our tiles to make the array and see. → If you multiply side lengths, you get area, so we can think about what numbers you can multiply to make 18.

T: Work with your partner to make a list of multiplication facts that equal 18.

S: (Possible list is as follows: 1×18, 18×1, 2×9, 9×2, 3×6, and 6×3.)

T: Let us draw a 3 cm by 6 cm rectangular array. Use a ruler to measure the side lengths on your personal white board. Mark each centimeter with a point and connect the points to draw the square centimeters.

T: After drawing your squares, check your work by skip-counting the rows to find the total number of tiles you drew.

S: (Draw, label, and skip-count tiles in array.)

T: Turn your personal white board so that it is vertical. Does the rectangle still have the same area?

S: Yes.

T: However, the side lengths switched places. Tell your partner how you know the area is the same.

S: The side lengths didn't change; they just moved. → It is the commutative property. We learned before that you can turn an array and it does not change how much is in it. The rows just turn into columns and columns turn into rows.

Part 3: Interpret area models to find area.

T: The grid you drew inside of your 3 cm by 6 cm rectangle shows a picture of all the tiles that make up the area. Carefully erase the grid lines in your rectangle. (Pause.) The empty rectangle with labeled side lengths left is called an **area model**. How can you find the total area just using the labeled side lengths?

S: I can multiply! → I can multiply the side lengths, 3 cm and 6 cm, to get the area, 18 square cm.

T: (Project or draw the area model on the right.) What is the total area of my pictured rectangle?

S: 18 square cm.

T: Tell your partner how you figured out the area.

S: It is easy. One side length is 18 and the other is 1. 18 × 1 = 18. The labels tell you the unit is centimeters, so the area is square centimeters.

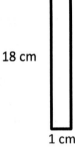

18 cm

1 cm

T: (Pass out the area model template.) Slip the area model into your personal white board. Use your ruler to measure the side lengths of one of the squares on the grid. (Allow students time to measure.) What unit makes up this grid?

S: Square inches!

T: The side lengths of this area model are not labeled. Let us draw a grid inside it to help find the side lengths. Earlier, we drew a grid inside a rectangle by marking each unit with a point and using a ruler to connect the points. Do we need to draw points on the area model to draw a grid inside of it?

S: No, we can just use the grid lines. → No, the lines on the grid can act as points because the area model is lined up with the grid.

Area Model Template

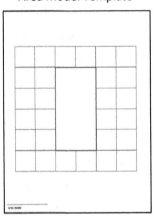

T: Use your ruler and the lines on the grid to draw squares inside of the area model. (Allow students time to work.) What size are the units inside the area model?

S: Square inches. → They are square inches because we used the square inch grid paper to help draw the squares.

T: Find and label the side lengths, and then write an equation to find the area.

S: (Label the side lengths as 4 in and 2 in, and write 2 × 4 = 8 or 4 × 2 = 8.)

T: What is the area?

S: 8 square inches!

EUREKA
MATH™

Problem Set (10 minutes)

Students should do their personal best to complete the Problem Set within the allotted 10 minutes. For some classes, it may be appropriate to modify the assignment by specifying which problems they work on first. Some problems do not specify a method for solving. Students should solve these problems using the RDW approach used for Application Problems.

Student Debrief (10 minutes)

Lesson Objective: Interpret area models to form rectangular arrays.

The Student Debrief is intended to invite reflection and active processing of the total lesson experience.

Invite students to review their solutions for the Problem Set. They should check work by comparing answers with a partner before going over answers as a class. Look for misconceptions or misunderstandings that can be addressed in the Debrief. Guide students in a conversation to debrief the Problem Set and process the lesson.

Any combination of the questions below may be used to lead the discussion.

- What was your strategy for finding the total number of squares in Problem 2(c)?

- Invite students who drew arrays that demonstrate commutativity for Problem 4(a) (possibly 4 × 6 and 6 × 4) to share their work. Guide students to articulate understanding that commutativity still applies in the context of area.

- For Problem 4(b), most students answered that Mrs. Barnes' array probably had 24 squares. Is there another answer that makes sense? (For example, 12, 48, 72.)

- Compare the **area model** to the array. How are they the same and different? (Guide discussion to include the commutativity of both models.)

NOTES ON MULTIPLE MEANS FOR ACTION AND EXPRESSION:

Consider offering the following adaptations to the Problem Set:

- Prompt students to approach Rectangle E first. Offer practice with 1 by *n* rectangles to build fluency and confidence.

- Remove side lengths to encourage closer investigation.

- Challenge students to devise an alternate method for finding the area of Benjamin's bedroom floor.

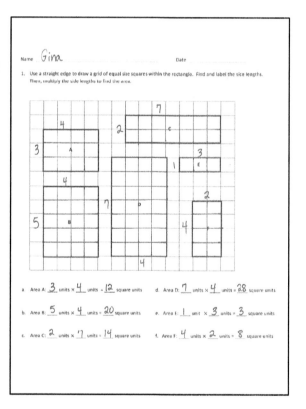

Exit Ticket (3 minutes)

After the Student Debrief, instruct students to complete the Exit Ticket. A review of their work will help with assessing students' understanding of the concepts that were presented in today's lesson and planning more effectively for future lessons. The questions may be read aloud to the students.

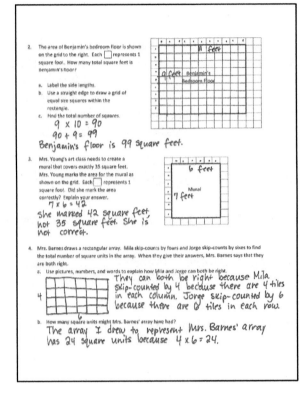

Name _____ Date _____

1. Use a straight edge to draw a grid of equal size squares within the rectangle. Find and label the side lengths. Then, multiply the side lengths to find the area.

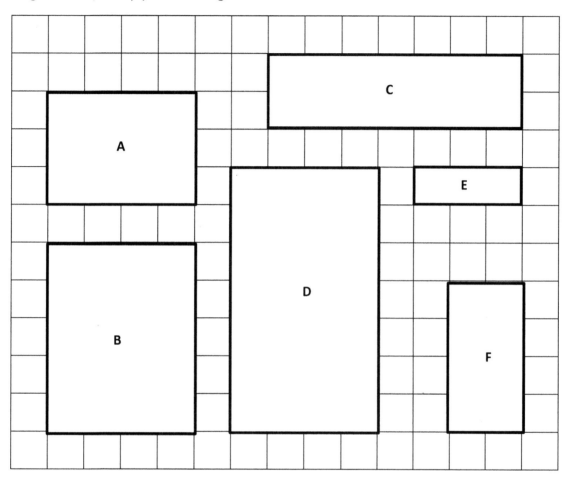

a. Area A:

    ____ units × ____ units = ____ square units

b. Area B:

    ____ units × ____ units = ____ square units

c. Area C:

    ____ units × ____ units = ____ square units

d. Area D:

    ____ units × ____ units = ____ square units

e. Area E:

    ____ unit × ____ units = ____ square units

f. Area F:

    ____ units × ____ units = ____ square units

©2015 Great Minds. eureka math.org
G3-M4-TE-B4-1.3.1-01.2016 -

2. The area of Benjamin's bedroom floor is shown on the grid to the right. Each ☐ represents 1 square foot. How many total square feet is Benjamin's floor?

   a. Label the side lengths.

   b. Use a straight edge to draw a grid of equal size squares within the rectangle.

   c. Find the total number of squares.

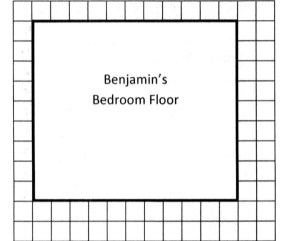

3. Mrs. Young's art class needs to create a mural that covers exactly 35 square feet. Mrs. Young marks the area for the mural as shown on the grid. Each ☐ represents 1 square foot. Did she mark the area correctly? Explain your answer.

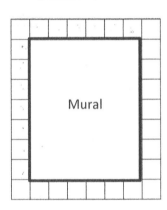

4. Mrs. Barnes draws a rectangular array. Mila skip-counts by fours and Jorge skip-counts by sixes to find the total number of square units in the array. When they give their answers, Mrs. Barnes says that they are both right.

   a. Use pictures, numbers, and words to explain how Mila and Jorge can both be right.

   b. How many square units might Mrs. Barnes' array have had?

      Lesson 7:   Interpret area models to form rectangular arrays.

©2015 Great Minds. eureka-math.org
G3-M4-TE-B4-1.3.1-01.2016

EUREKA MATH™

Name _____   Date _____

1. Label the side lengths of Rectangle A on the grid below. Use a straight edge to draw a grid of equal size squares within Rectangle A. Find the total area of Rectangle A.

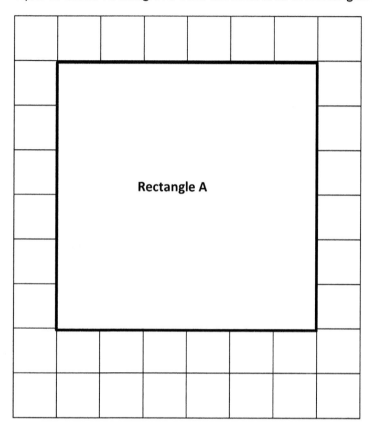

Rectangle A

Area: _____ square units

2. Mark makes a rectangle with 36 square centimeter tiles. Gia makes a rectangle with 36 square inch tiles. Whose rectangle has a bigger area? Explain your answer.

Name _____ Date _____

1. Find the area of each rectangular array. Label the side lengths of the matching area model, and write a multiplication equation for each area model.

Rectangular Arrays	Area Models
a. 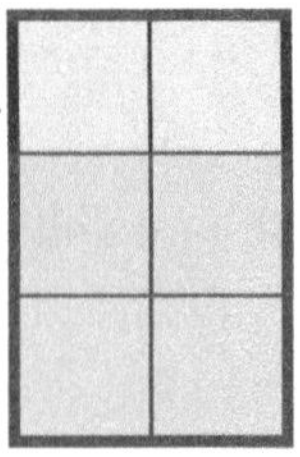 _____ square units	3 units  3 units × _____ units = _____ square units 2 units
b. 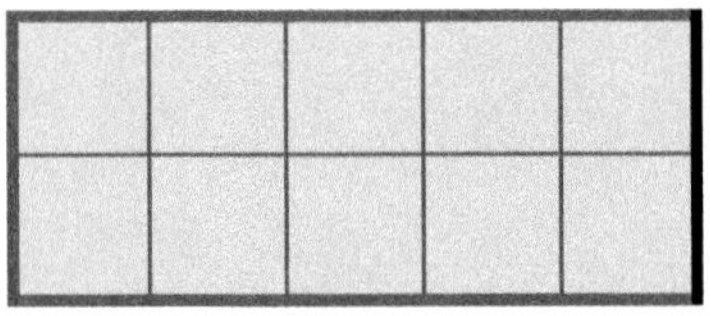 _____ square units	 _____ units × _____ units = _____ square units
c. 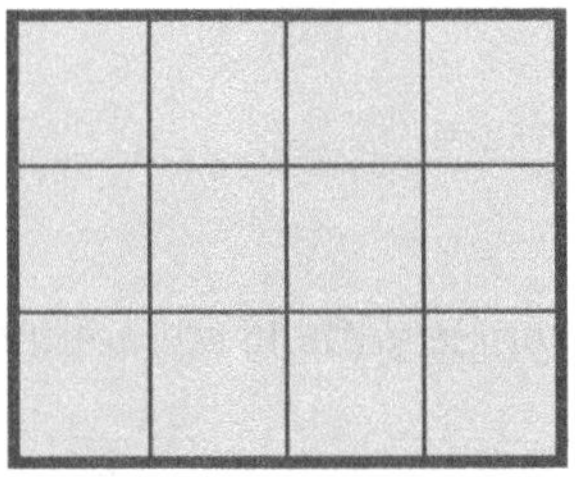 _____ square units	_____ units × _____ units = _____ square units
d. 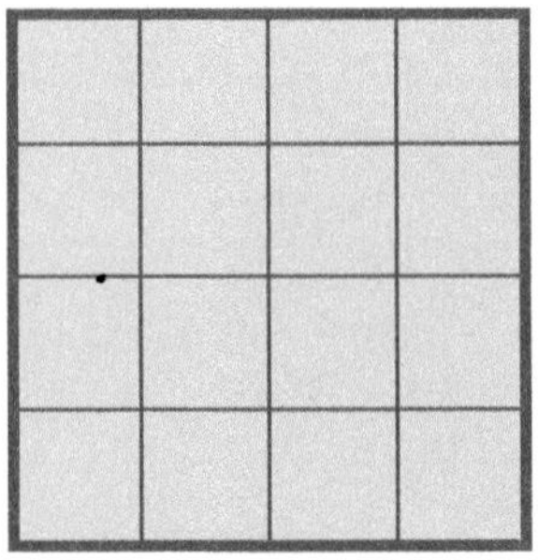 _____ square units	_____ units × _____ units = _____ square units

©2015 Great Minds. eureka-math.org
G3-M4-TE-B4-1.3.1-01.2016

EUREKA
MATH™

2. Jillian arranges square pattern blocks into a 7 by 4 array. Draw Jillian's array on the the grid below. How many square units are in Jillian's rectangular array?

a.

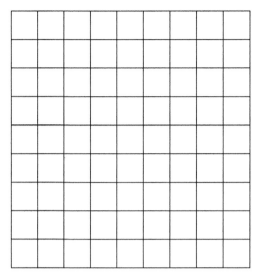

b. Label the side lengths of Jillian's array from Part (a) on the rectangle below. Then, write a multiplication sentence to represent the area of the rectangle.

3. Fiona draws a 24 square centimeter rectangle. Gregory draws a 24 square inch rectangle. Whose rectangle is larger in area? How do you know?

area model

Lesson 7: Interpret area models to form rectangular arrays.

EUREKA MATH

Lesson 8

Objective: Find the area of a rectangle through multiplication of the side lengths.

Suggested Lesson Structure

■ Fluency Practice   (11 minutes)

░ Application Problem   (5 minutes)

░ Concept Development   (34 minutes)

■ Student Debrief   (10 minutes)

Total Time   **(60 minutes)**

Fluency Practice (11 minutes)

- Multiply by 6 **3.OA.7**   (8 minutes)
- Group Counting **3.OA.1**   (3 minutes)

Multiply by 6 (8 minutes)

Materials: (S) Multiply by 6 (6–10) Pattern Sheet

Note: This activity builds fluency with respect to multiplication facts using units of 6. It works toward students knowing from memory all products of two one-digit numbers. See Lesson 2 for the directions for administration of a Multiply-By Pattern Sheet.

  T: (Write $7 \times 6 =$ ____.) Let us skip-count up by sixes. (Count with fingers to 7 as students count.)

  S: 6, 12, 18, 24, 30, 36, 42.

  T: Let us see how we can skip-count down to find the answer, too. (Show 10 fingers.) Start at 60. (Count down with your fingers as students say numbers.)

  S: 60, 54, 48, 42.

Continue with the following possible sequence: 9×6, 6×6, and 8×6.

  T: (Distribute Multiply by 6 (6–10) Pattern Sheet.) Let us practice multiplying by 6. Be sure to work left to right across the page.

©2015 Great Minds. eureka math.org
G3-M4-TE-B4-1.3.1-01.2016 -

Group Counting (3 minutes)

Note: Group counting reviews interpreting multiplication as repeated addition.

Instruct students to count forward and backward, occasionally changing the direction of the count.

- Fours to 40
- Sevens to 70
- Eights to 80
- Nines to 90

Application Problem (5 minutes)

Marnie and Connor both skip-count square units to find the area of the same rectangle. Marnie counts, "3, 6, 9, 12, 15, 18, 21." Connor counts, "7, 14, 21." Draw what the rectangle might look like, and then label the side lengths and find the area.

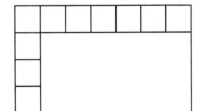

$$3 \times 7 = 21$$

The area is 21 square units.

Note: This problem reinforces Lesson 7 and sets the foundation for today's Concept Development. Invite students to share their drawings and discuss how they are similar and different.

Concept Development (34 minutes)

Materials: (S) Personal white board, inch ruler, grid (Template)

Grid Template

Part 1: Relate side lengths to area.

T: (Project image shown below.) How many rows are in the incomplete array?

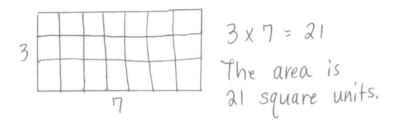

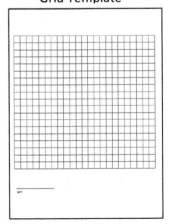

Lesson 8:   Find the area of a rectangle through multiplication of the side lengths.

EUREKA
MATH™

©2015 Great Minds. eureka-math.org
G3-M4-TE-B4-1.3.1-01.2016

S: 4 rows.

T: How many square units are there in each row?

S: 7 square units.

T: Talk to your partner: Do we need to complete the array to find the area of the rectangle? Why or why not?

S: Yes, then we can skip-count each row to find the total. → No, we already know the side lengths!

T: How are the side lengths related to the area?

S: If you multiply the side lengths together, the product is the same as the area.

MP.8

T: Talk to a partner: Can you multiply any two side lengths to find the area?

S: No, you have to multiply the side length that shows the number of rows times the side length that shows the number of squares in each row.

T: What multiplication equation can be used to find the area of this rectangle?

S: $4 \times 7 = 28$.

T: To check our answer, use your grid template to trace and shade in an area model that is 4 units high and 7 units wide. Label each side length.

S: (Draw and label.)

T: Was our answer correct?

S: Yes, I used the grid paper to count 28 squares inside. → I skip-counted 4 sevens to get 28.

Continue with the following possible sequence: 6×5, 8×7, and 9×6.

Part 2: Use side lengths to find area.

Draw or project the rectangle shown on the right.

T: What do you notice about this rectangle?

S: We know the side lengths, but there is no grid inside. → It is an area model.

T: Do we still have enough information to find the area of this rectangle, even without the grid lines inside?

S: Yes! We know both side lengths.

T: Write the multiplication equation to find the area of this rectangle.

S: $6 \times 8 = 48$.

You may want to help English language learners relate the number of square units in each row to the word *columns* and relate *columns* and *rows* to side lengths. To some students, it may appear that these words are used interchangeably. Help clarify their meaning.

NOTES ON MULTIPLE MEANS OF ENGAGEMENT:

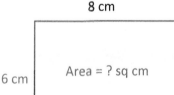

Lesson 8:   Find the area of a rectangle through multiplication of the side lengths.

93

©2015 Great Minds. eureka math.org
G3-M4-TE-B4-1.3.1-01.2016 -

Continue with the following suggested examples, having students work independently or in pairs:

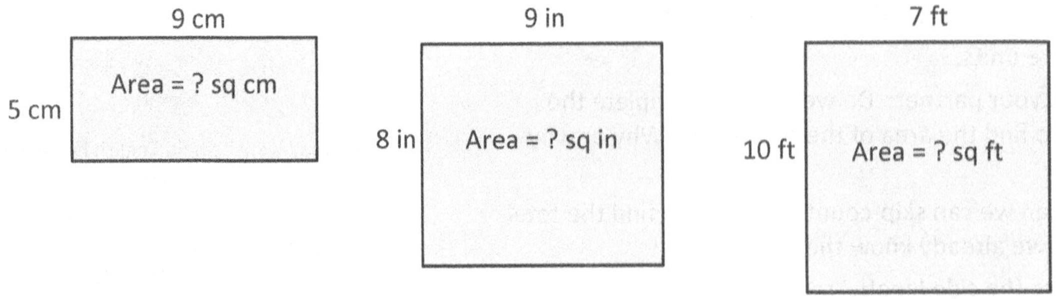

9 cm

5 cm | Area = ? sq cm |

9 in

8 in | Area = ? sq in |

7 ft

10 ft | Area = ? sq ft |

Part 3: Use area and side length to find unknown side length.

Draw or project the rectangle shown on the right.

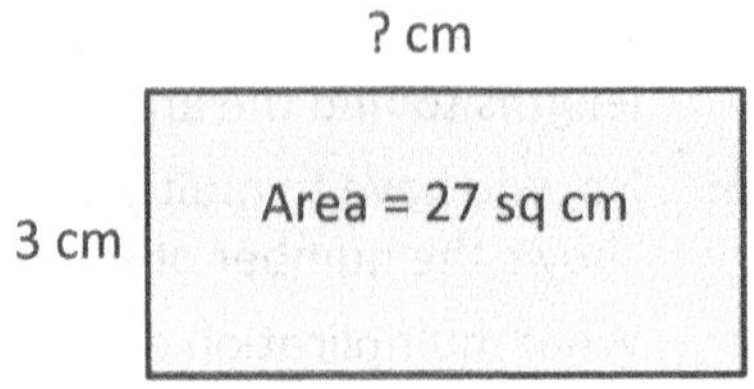

T: What do you notice about this rectangle?

S: We know the area but not both side lengths.
→ One of the side lengths is unknown.

T: Write a multiplication equation on your personal white board to show how to find the area of this rectangle. Use a question mark for the unknown side length.

S: (Write 3 × ? = 27.)

T: What is the value of the question mark?

S: 9.

T: How do you know?

S: I know that 3 times 9 equals 27.

T: So, what is the unknown side length?

S: 9 centimeters!

T: Write the related division equation on your board.

S: (Write 27 ÷ 3 = 9.)

? cm

3 cm | Area = 27 sq cm |

Continue with the following suggested examples:

? cm

4 cm | Area = 32 sq cm |

? ft

8 ft | Area = 24 sq ft |

7 in

? in | Area = 42 sq in |

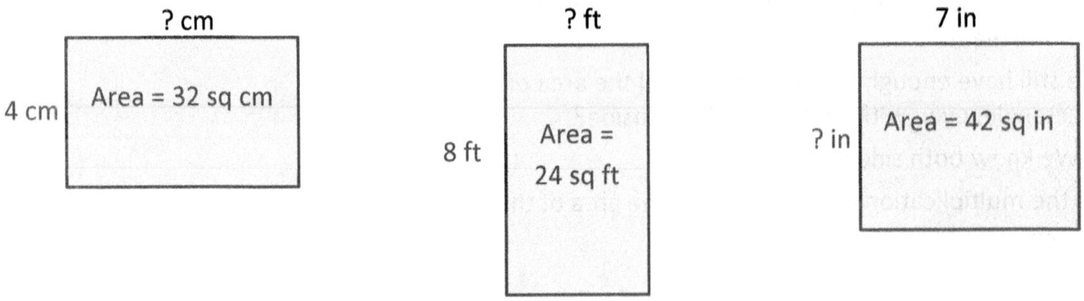

T: When you know the area and one side length of a rectangle, how can you find the other side length?

S: I can think of it as a multiplication equation with an unknown factor. → Or, I can divide the area by the known side length.

Lesson 8: Find the area of a rectangle through multiplication of the side lengths.

EUREKA
MATH™

Problem Set (10 minutes)

Students should do their personal best to complete the Problem Set within the allotted 10 minutes. For some classes, it may be appropriate to modify the assignment by specifying which problems they work on first. Some problems do not specify a method for solving. Students should solve these problems using the RDW approach used for Application Problems.

Student Debrief (10 minutes)

Lesson Objective: Find the area of a rectangle through multiplication of the side lengths.

The Student Debrief is intended to invite reflection and active processing of the total lesson experience.

Invite students to review their solutions for the Problem Set. They should check work by comparing answers with a partner before going over answers as a class. Look for misconceptions or misunderstandings that can be addressed in the Debrief. Guide students in a conversation to debrief the Problem Set and process the lesson.

Any combination of the questions below may be used to lead the discussion.

- In what way is the area of Problem 1(b) related to the area of Problem 1(a)? (It is double.) How could you use the side lengths to help you figure out that 8 × 7 is double 4 × 7?

- Which shape in Problem 1 is a square? How do you know?

- How are the rectangles in Problem 1(a) and 2(c) similar? How are they different?

- Address the following possible misconception in Problem 5. Although Eliza's bedroom has 1 side length (6 feet) that is 1 more than her brother's bedroom (5 feet), and 1 side length (7 feet) that is 1 less than her brother's bedroom (8 feet), the floor areas are not equal.

- Why is there a connection between a rectangle's side lengths and its area?

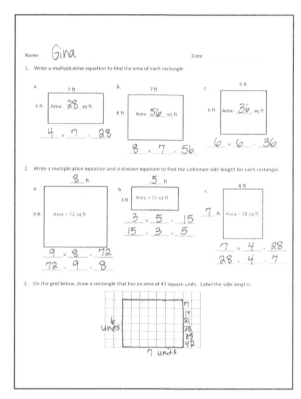

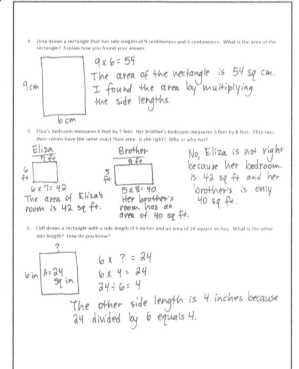

Lesson 8:   Find the area of a rectangle through multiplication of the side lengths.

©2015 Great Minds. eureka-math.org
G3-M4-TE-B4-1.3.1-01.2016 -

95

Exit Ticket (3 minutes)

After the Student Debrief, instruct students to complete the Exit Ticket. A review of their work will help with assessing students' understanding of the concepts that were presented in today's lesson and planning more effectively for future lessons. The questions may be read aloud to the students.

Lesson 8: Find the area of a rectangle through multiplication of the side lengths.

©2015 Great Minds. eureka-math.org
G3-M4-TE-B4-1.3.1-01.2016

Multiply.

6 x 1 = _____   6 x 2 = _____   6 x 3 = _____   6 x 4 = _____

6 x 5 = _____   6 x 6 = _____   6 x 7 = _____   6 x 8 = _____

6 x 9 = _____   6 x 10 = _____   6 x 5 = _____   6 x 6 = _____

6 x 5 = _____   6 x 7 = _____   6 x 5 = _____   6 x 8 = _____

6 x 5 = _____   6 x 9 = _____   6 x 5 = _____   6 x 10 = _____

6 x 6 = _____   6 x 5 = _____   6 x 6 = _____   6 x 7 = _____

6 x 6 = _____   6 x 8 = _____   6 x 6 = _____   6 x 9 = _____

6 x 6 = _____   6 x 7 = _____   6 x 6 = _____   6 x 7 = _____

6 x 8 = _____   6 x 7 = _____   6 x 9 = _____   6 x 7 = _____

6 x 8 = _____   6 x 6 = _____   6 x 8 = _____   6 x 7 = _____

6 x 8 = _____   6 x 9 = _____   6 x 9 = _____   6 x 6 = _____

6 x 9 = _____   6 x 7 = _____   6 x 9 = _____   6 x 8 = _____

6 x 9 = _____   6 x 8 = _____   6 x 6 = _____   6 x 9 = _____

6 x 7 = _____   6 x 9 = _____   6 x 6 = _____   6 x 8 = _____

6 x 9 = _____   6 x 7 = _____   6 x 6 = _____   6 x 8 = _____

multiply by 6 (6–10)

©2015 Great Minds. eureka math.org
G3-M4-TE-B4-1.3.1-01.2016 -

Name _____   Date _____

1. Write a multiplication equation to find the area of each rectangle.

a.
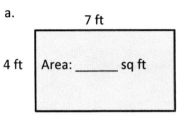
7 ft

4 ft   Area: _____ sq ft

_____ × _____ = _____

b.
7 ft

8 ft   Area: _____ sq ft

_____ × _____ = _____

c.
6 ft
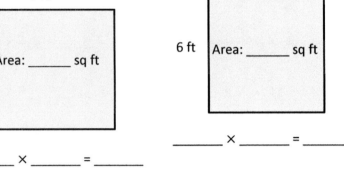
6 ft   Area: _____ sq ft

_____ × _____ = _____

2. Write a multiplication equation and a division equation to find the unknown side length for each rectangle.

a.
_____ ft
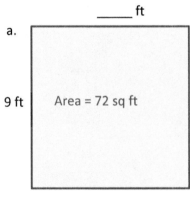
9 ft   Area = 72 sq ft

_____ × _____ = _____

_____ ÷ _____ = _____

b.
_____ ft
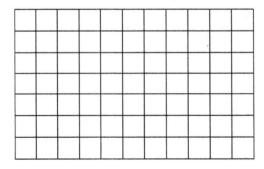
3 ft   Area = 15 sq ft

_____ × _____ = _____

_____ ÷ _____ = _____

c.
4 ft

_____ ft   Area = 28 sq ft

_____ × _____ = _____

_____ ÷ _____ = _____

3. On the grid below, draw a rectangle that has an area of 42 square units. Label the side lengths.

Lesson 8:   Find the area of a rectangle through multiplication of the side lengths.

EUREKA
MATH™

4. Ursa draws a rectangle that has side lengths of 9 centimeters and 6 centimeters. What is the area of the rectangle? Explain how you found your answer.

5. Eliza's bedroom measures 6 feet by 7 feet. Her brother's bedroom measures 5 feet by 8 feet. Eliza says their rooms have the same exact floor area. Is she right? Why or why not?

6. Cliff draws a rectangle with a side length of 6 inches and an area of 24 square inches. What is the other side length? How do you know?

EUREKA
MATH

Lesson 8:   Find the area of a rectangle through multiplication of the side lengths.

99

©2015 Great Minds. eureka math.org
G3-M4-TE-B4-1.3.1-01.2016 -

Name _____   Date _____

1. Write a multiplication equation to find the area of the rectangle below.

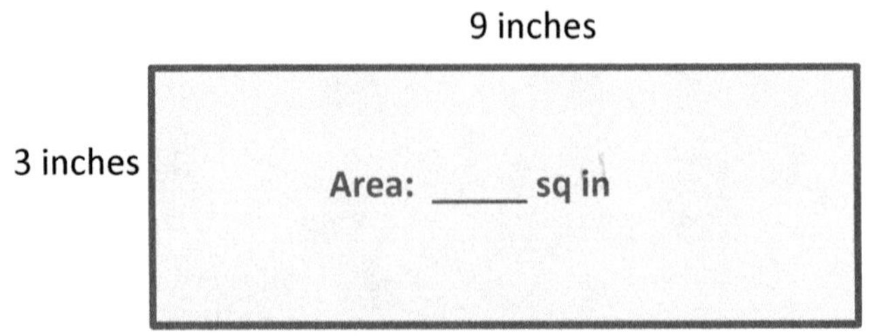

9 inches

3 inches

Area: _____ sq in

_____ × _____ = _____

2. Write a multiplication equation and a division equation to find the unknown side length for the rectangle below.

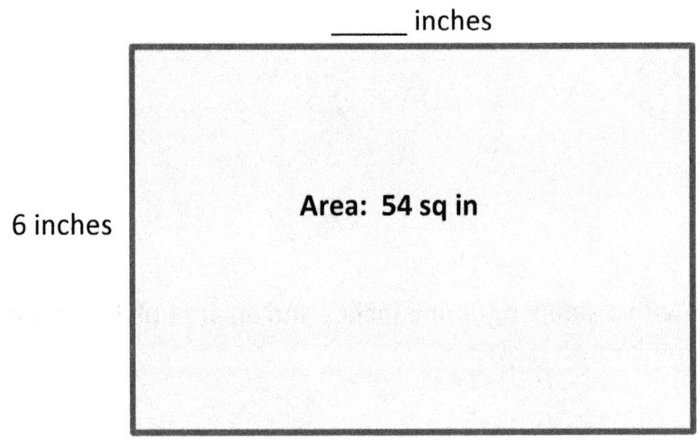

_____ inches

6 inches

Area: 54 sq in

_____ × _____ = _____

_____ ÷ _____ = _____

Lesson 8:   Find the area of a rectangle through multiplication of the side lengths.

EUREKA
MATH™

Name _____ Date _____

1. Write a multiplication equation to find the area of each rectangle.

a.

8 cm

3 cm   Area: _____ sq cm

_____ × _____ = _____

b.

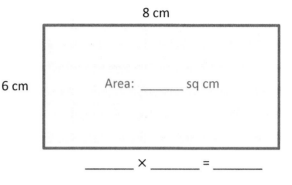

8 cm

6 cm   Area: _____ sq cm

_____ × _____ = _____

c.

4 ft

4 ft   Area: _____ sq ft

_____ × _____ = _____

d.

7 ft

4 ft   Area: _____ sq ft

_____ × _____ = _____

2. Write a multiplication equation and a division equation to find the unknown side length for each rectangle.

a.

_____ ft.

3 ft   Area: 24 sq ft

_____ × _____ = _____

_____ ÷ _____ = _____

b.

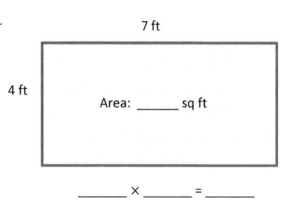

9 ft

_____ ft   Area: 36 sq ft

_____ × _____ = _____

_____ ÷ _____ = _____

EUREKA
MATH™

Lesson 8:   Find the area of a rectangle through multiplication of the side lengths.

101

©2015 Great Minds. eureka math.org
G3-M4-TE-B4-1.3.1-01.2016 -

3. On the grid below, draw a rectangle that has an area of 32 square centimeters. Label the side lengths.

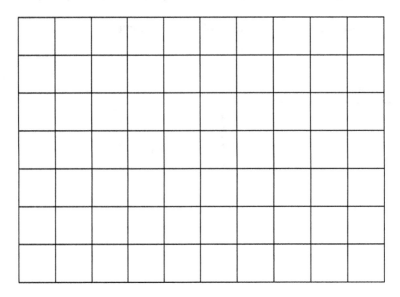

4. Patricia draws a rectangle that has side lengths of 4 centimeters and 9 centimeters. What is the area of the rectangle? Explain how you found your answer.

5. Charles draws a rectangle with a side length of 9 inches and an area of 27 square inches. What is the other side length? How do you know?

Lesson 8:   Find the area of a rectangle through multiplication of the side lengths.

EUREKA
MATH™

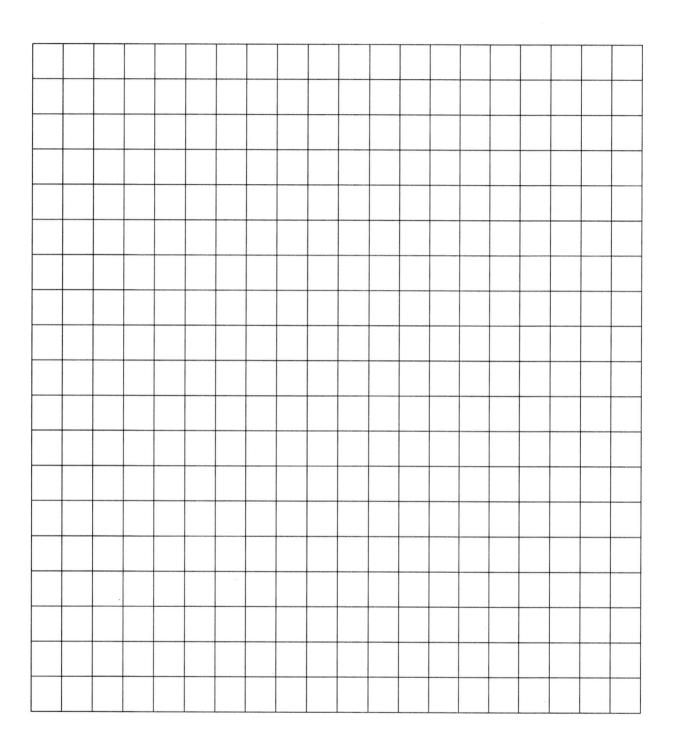

grid

Lesson 8: Find the area of a rectangle through multiplication of the side lengths.

103

©2015 Great Minds. eureka math.org
G3-M4-TE-B4-1.3.1-01.2016

Name _____ Date _____

1. Jasmine and Roland each use unit squares to tile a piece of paper. Their work is shown below.

Jasmine's Array

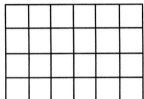

Roland's Array

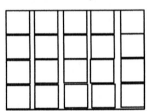

a. Can one of the arrays be used to correctly measure the area of the piece of paper? If so, whose array would you use? Explain why.

b. What is the area of the piece of paper? Explain your strategy for finding the area.

c. Jasmine thinks she can skip-count by sixes to find the area of her rectangle. Is she correct? Explain why or why not.

©2015 Great Minds. eureka-math.org
G3-M4-TE-B4-1.3.1-01.2016

EUREKA
MATH™

2. Jaheim says you can create three rectangles with different side lengths using
    12 unit squares. Use pictures, numbers, and words to show what Jaheim is saying.

3. The area of a rectangle is 72 square units. One side has a length of 9 units. What is the other side length?
    Explain how you know using pictures, equations, and words.

4. Jax started to draw a grid inside the rectangle to find its area.

    a. Use a straight edge to complete the drawing of the grid.

    b. Write a skip-count sequence you could use to find the area.

    c. Write a multiplication equation that you could use to find the area, and then solve.

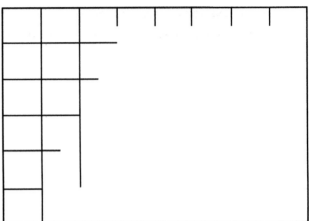

5. Half of the rectangle below has been tiled with unit squares.

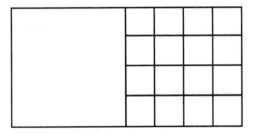

    a. How many more unit squares are needed to fill in the rest of the rectangle?

    b. What is the total area of the large rectangle? Explain how you found the area.

©2015 Great Minds. eureka-math.org
G3-M4-TE-B4-1.3.1-01.2016

EUREKA
MATH™

Geometric measurement: understand concepts of area and relate area to multiplication and to addition.

3.MD.5   Recognize area as an attribute of plane figures and understand concepts of area measurement.

      a.   A square with side length 1 unit, called "a unit square," is said to have "one square unit" of area, and can be used to measure area.

      b.   A plane figure which can be covered without gaps or overlaps by *n* unit squares is said to have an area of *n* square units.

3.MD.6   Measure areas by counting unit squares (square cm, square m, square in, square ft, and improvised units).

3.MD.7   Relate area to the operations of multiplication and addition.

      a.   Find the area of a rectangle with whole-number side lengths by tiling it, and show that the area is the same as would be found by multiplying the side lengths.

      b.   Multiply side lengths to find areas of rectangles with whole-number side lengths in the context of solving real world and mathematical problems, and represent whole-number products as rectangular areas in mathematical reasoning.

      d.   Recognize area as additive. Find areas of rectilinear figures by decomposing them into non-overlapping rectangles and adding the areas of the non-overlapping parts, applying this technique to solve real world problems.

Evaluating Student Learning Outcomes

A Progression Toward Mastery is provided to describe steps that illuminate the gradually increasing understandings that students develop on their way to proficiency. In this chart, this progress is presented from left (Step 1) to right (Step 4). The learning goal for students is to achieve Step 4 mastery. These steps are meant to help teachers and students identify and celebrate what the students CAN do now and what they need to work on next.

©2015 Great Minds. eureka math.org
G3-M4-TE-B4-1.3.1-01.2016 -

A Progression Toward Mastery				
Assessment Task Item and Standards Assessed	STEP 1 Little evidence of reasoning without a correct answer. (1 Point)	STEP 2 Evidence of some reasoning without a correct answer. (2 Points)	STEP 3 Evidence of some reasoning with a correct answer or evidence of solid reasoning with an incorrect answer. (3 Points)	STEP 4 Evidence of solid reasoning with a correct answer. (4 Points)
1 3.MD.5 3.MD.6	Response demonstrates little evidence of reasoning without a correct answer.	Response shows limited reasoning with at least one correct answer.	Response includes evidence of some reasoning with three correct answers or evidence of solid reasoning with an incorrect answer.	Student correctly answers: a. Jasmine's array, giving strong evidence of understanding that tiling must have no gaps or overlaps. b. The area is 24 square units. Student provides an appropriate explanation of the calculation (e.g., counting or skip-counting strategies). c. Yes, there are 4 rows of 6 squares, so it is possible to skip-count by six.
2 3.MD.7b	Response demonstrates little evidence of reasoning without a correct answer.	Response shows limited reasoning with at least one correct answer.	Student identifies at least two of three rectangles correctly. Response includes evidence of accurate reasoning with pictures, numbers, or words.	Student correctly identifies three rectangles: ▪ 1×12 or 12×1 ▪ 2×6 or 6×2 ▪ 3×4 or 4×3 Response shows evidence of solid reasoning using pictures, numbers, and words.

Module 4:   Multiplication and Area

©2015 Great Minds. eureka-math.org
G3-M4-TE-B4-1.3.1-01.2016

EUREKA MATH

A Progression Toward Mastery				
3 **3.MD.7b**	Response demonstrates little evidence of reasoning without a correct answer.	Response shows limited reasoning without a correct answer.	Student finds the unknown side length of 8 units but may not show enough work to clearly justify the answer.	Student correctly finds the unknown side length of 8 units. Response shows evidence of solid reasoning using pictures, equations, and words.
4 **3.MD.5** **3.MD.6** **3.MD.7a**	Response demonstrates little evidence of reasoning without a correct answer.	Response shows evidence of some reasoning in an attempt to write a skip-counting sequence, an equation and complete the array, but work may not include a correct answer.	Student accurately completes the array and finds the area of 48 sq units but may not accurately provide both a skip-count sequence and a multiplication equation.	Student correctly: a. Completes the array with 8 columns and 6 rows. b. Writes one of the following skip-count sequences: 6, 12, 18, 24, 30, 36, 42, 48 OR 8, 16, 24, 32, 40, 48. c. Writes a multiplication equation ($6 \times 8 = 48$ or $8 \times 6 = 48$), and gives an area of 48 sq units.
5 **3.MD.5a** **3.MD.5b** **3.MD.7a** **3.MD.7d**	Response demonstrates little evidence of reasoning without a correct answer to either part.	Response shows limited reasoning with a correct answer in one part.	Student slightly miscalculates the number of tiles needed to fill the remaining area, but the explanation shows evidence of solid reasoning. Part (b) is correct based on the student's slight miscalculation but not the correct answer of 32 square units.	Student correctly: a. Identifies that 16 square units are needed to fill the remaining area. b. Says the area of the large rectangle is 32 square units. Explanation gives evidence of solid reasoning to support answer.

Name _Gina_

Date _____

1. Jasmine and Roland each use unit squares to tile a piece of paper. Their work is shown below.

Jasmine's Array

Roland's Array

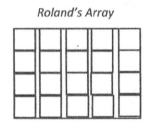

4 units

6 units

a. Can one of the arrays be used to correctly measure the area of the piece of paper? If so, whose array would you use? Explain why.

Jasmine's array correctly measures the area of the piece of paper. You can't have gaps or overlaps when you tile or it won't be right.

b. What is the area of the piece of paper? Explain your strategy for finding the area.

Area = 4 units × 6 units
       = 24 sq units

I multiplied the side lengths, 4 units × 6 units, to get the area, 24 sq units.

c. Jasmine thinks she can skip-count by sixes to find the area of her rectangle. Is she correct? Explain why or why not.

Yes, Jasmine is correct. There are 4 rows of six unit squares, so she can skip-count: 6, 12, 18, 24. It's faster if she multiplies.

110

Module 4:   Multiplication and Area

©2015 Great Minds. eureka-math.org
G3-M4-TE-B4-1.3.1-01.2016

EUREKA
MATH

2. Jaheim says you can create three rectangles with different side lengths using 12 unit squares. Use pictures, numbers, and words to show what Jaheim is saying.

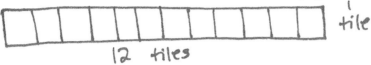

12 tiles | tile

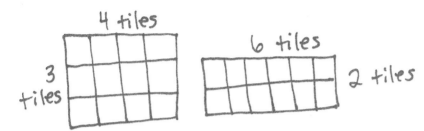

4 tiles
3 tiles
6 tiles
2 tiles

Jaheim is correct. These are the only rectangles you can make with 12 tiles. You can turn them, but they will be the same:

6 = 2

3. The area of a rectangle is 72 square units. One side has a length of 9 units. What is the other side length? Explain how you know using pictures, equations, and words.

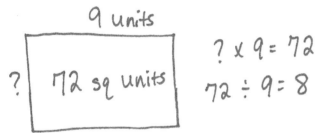

9 units
?   72 sq units

? × 9 = 72
72 ÷ 9 = 8

If one side length is 9 units, the other side length is 8 units because 8 × 9 = 72.

4. Jax started to draw a grid inside the rectangle to find its area.

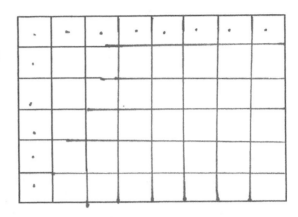

   a. Use a straight edge to complete the drawing of the grid.

   b. Write a skip-count sequence you could use to find the area.

   8, 16, 24, 32, 40, 48

        48 sq units

   c. Write a multiplication equation that you could use to find the area, and then solve.

   8 units × 6 units = 48 sq units

5. Half of the rectangle below has been tiled with unit squares.

   4 units

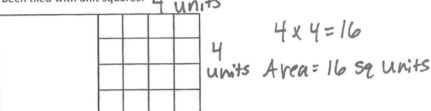

   4 units $4 \times 4 = 16$

   Area = 16 sq units

   a. How many more unit squares are needed to fill in the rest of the rectangle?

   If there are 16 sq units in one half, there will be 16 sq units in the other half too. You need 16 more unit squares to fill it in.

   b. What is the total area of the large rectangle? Explain how you found the area.

   16 sq units + 16 sq units = 32 sq units
   I added the 2 halves together to find the total area.

EUREKA MATH™

Mathematics Curriculum

GRADE 3

Topic C
Arithmetic Properties Using Area Models

3.MD.5, 3.MD.7

Focus Standards:	3.MD.5	Recognize area as an attribute of plane figures and understand concepts of area measurement. a. A square with side length 1 unit, called "a unit square," is said to have "one square unit" of area, and can be used to measure area. b. A plane figure which can be covered without gaps or overlaps by *n* unit squares is said to have an area of *n* square units.
	3.MD.7	Relate area to the operations of multiplication and addition. a. Find the area of a rectangle with whole-number side lengths by tiling it, and show that the area is the same as would be found by multiplying the side lengths. b. Multiply side lengths to find areas of rectangles with whole-number side lengths in the context of solving real world and mathematical problems, and represent whole-number products as rectangular areas in mathematical reasoning. c. Use tiling to show in a concrete case that the area of a rectangle with whole-number side lengths *a* and *b* + *c* is the sum of *a* × *b* and *a* × *c*. Use area models to represent the distributive property in mathematical reasoning. d. Recognize area as additive. Find areas of rectilinear figures by decomposing them into non-overlapping rectangles and adding the areas of the non-overlapping parts, applying this technique to solve real world problems.
Instructional Days:	3	
Coherence -Links from:	G2–M2	Addition and Subtraction of Length Units
	G3–M1	Properties of Multiplication and Division and Solving Problems with Units of 2–5 and 10
	G3–M3	Multiplication and Division with Units of 0, 1, 6–9, and Multiples of 10
-Links to:	G4–M3	Multi-Digit Multiplication and Division
	G4–M7	Exploring Multiplication

Topic C begins with a concrete study of arithmetic properties. Students cut apart rectangular grids and rearrange the parts to create new rectangles with the same area. Lesson 9 lays the foundation for the work to come in Lessons 10 and 11.

In Lesson 10, students apply knowledge of the distributive property from Modules 1 and 3 to find area. In previous modules, they learned to decompose an array of discrete items into two parts, determine the number of units in each part, and then find the sum of the parts. Now, students connect this experience to using the distributive property to determine the unknown side length of an array that may, for example, have an area of 72 square units. They might decompose the area into an 8 by 5 rectangle and an 8 by 4 rectangle. The sum of the side lengths, 5 + 4, gives the length of the unknown side.

In Lesson 11, students use a given number of square units to determine all possible whole number side lengths of rectangles with that area. Students engage in MP.3 as they justify that they have found all possible solutions for each given area using the associative property. Areas of 24, 36, 48, and 72 are chosen to reinforce multiplication facts that are often more difficult. Students realize that different factors give the same product. For example, they find that 4 by 12, 6 by 8, 1 by 48, and 2 by 24 arrays all have an area of 48 square units. They use understanding of the commutative property to recognize that area models can be rotated similar to the arrays in Modules 1 and 3.

A Teaching Sequence Toward Mastery of Arithmetic Properties Using Area Models
Objective 1: Analyze different rectangles and reason about their area. (Lesson 9)
Objective 2: Apply the distributive property as a strategy to find the total area of a larger rectangle by adding two products. (Lesson 10)
Objective 3: Demonstrate the possible whole number side lengths of rectangles with areas of 24, 36, 48, or 72 square units using the associative property. (Lesson 11)

EUREKA
MATH™

Lesson 9

Objective: Analyze different rectangles and reason about their area.

Suggested Lesson Structure

■ Fluency Practice   (12 minutes)
▨ Application Problem   (5 minutes)
▢ Concept Development   (33 minutes)
■ Student Debrief   (10 minutes)

   Total Time   **(60 minutes)**

Fluency Practice (12 minutes)

- Group Counting **3.OA.1**   (4 minutes)
- Find the Area **3.MD.7**   (4 minutes)
- Decompose the Multiplication Equation **3.OA.5**   (4 minutes)

Group Counting (4 minutes)

Note: Group counting reviews interpreting multiplication as repeated addition.

Instruct students to count forward and backward, occasionally changing the direction of the count.

- Fours to 40
- Sevens to 70
- Eights to 80
- Nines to 90

Find the Area (4 minutes)

Note: This fluency activity reviews strategies for finding the area of a rectangle.

   T: (Project a rectangular array with 2 rows of 4 units. Write 1 tile = 1 square meter.) What does 1 tile equal?
   S: 1 square meter.
   T: (Point to the side length of 4 units.) What is the value of this side length?
   S: 4 meters.
   T: (Point to the side length of 2 units.) What is the value of this side length?
   S: 2 meters.

Lesson 9:   Analyze different rectangles and reason about their area.

©2015 Great Minds. eureka math.org
G3-M4-TE-B4-1.3.1-01.2016 -

115

T: Write a multiplication sentence to represent the area of the rectangle.

S: (Write 2 m × 4 m = 8 sq m or 4 m × 2 m = 8 sq m.)

Continue with the following possible sequence: 3 rows of 5 units, 3 rows of 7 units, 4 rows of 6 units, 4 rows of 9 units, and 6 rows of 8 units.

Decompose the Multiplication Equation (4 minutes)

Materials: (S) Personal white board

Note: This activity anticipates the distributive property used in Lesson 10, while reviewing Module 3 concepts.

T: (Write 8 × 6 = (5 + ___) × 6.) Copy the equation on your personal white board, and fill in the blank.

S: (Write 8 × 6 = (5 + 3) × 6.)

T: (Write = (___ × 6) + (___ × 6).) Copy the equation on your personal white board, and fill in the blanks.

S: (Write (5 × 6) + (3 × 6).)

T: Solve the multiplication problems and write an addition equation. Below it, write your answer.

S: (Write 30 + 18 and 48 below it.)

Sample Work

8 × 6	= (5 + 3) × 6
	= (5 × 6) + (3 × 6)
	= 30 + 18
	= 48

Continue with the following possible sequence: 7 × 6, 6 × 6, and 9 × 6.

Application Problem (5 minutes)

Mario plans to completely cover his 8-inch by 6-inch piece of cardboard with square inch tiles. He has 42 square inch tiles. How many more square inch tiles does Mario need to cover the cardboard without any gaps or overlap? Explain your answer.

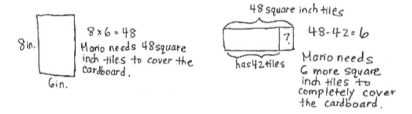

Note: This problem reviews the concept of finding area. Students will likely solve by multiplying side lengths (shown above), having just practiced this strategy in Lesson 8.

EUREKA
MATH

Concept Development (33 minutes)

Materials: (S) Small centimeter grid (Template), personal white board, Problem Set

Problems 1 and 2 in the Problem Set

- T: How can we cut this centimeter grid to get 2 equal rectangles?
- S: Cut it in half. → If we cut on the line between the fifth and sixth rows, we'll have 2 equal rectangles. → If we fold the grid in half and cut along the fold, we can make 2 equal rectangles.
- T: Do that now, and then answer Problem 1(a).

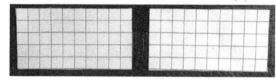

- T: (Point to the side length of 10 centimeters.) This is the **length** of the rectangle. What is its value?
- S: 10 centimeters.
- T: (Point to the side length of 5 centimeters.) This is the **width** of the rectangle. What is its value?
- S: 5 centimeters.
- T: How can you find the area of one of the rectangles?
- S: Multiply the side lengths. → Multiply 5 times 10.
- T: Answer Problem 1(b). (Allow students time to work.) What is the area of one of the rectangles?
- S: 50 square centimeters!
- T: What is the area of the other rectangle? How do you know?
- S: 50 square centimeters because the rectangles are equal.
- T: How can you find the total area of the rectangles?
- S: Add 50 square centimeters plus 50 square centimeters.
- T: Answer Problem 1(c). (Allow students time to work.) What is the total area?
- S: 100 square centimeters.
- T: Place your rectangles next to each other to make 1 long rectangle. Talk to a partner. What do you think the area of this long rectangle is? Why?

NOTES ON
MULTIPLE MEANS
OF ACTION AND
EXPRESSION:

Cutting paper with scissors may be a challenge for some learners. Precision is important to this lesson. Please try the following tips:

- Provide centimeter grids on cardstock or thicker paper.
- Darken and thicken the cutting lines.
- Provide left-handed, loop, spring, self-opening, or other adaptive scissors, if needed.
- Instruct students to turn the paper, not the scissors.
- Offer precut centimeter grids.

Small Centimeter
Grid Template

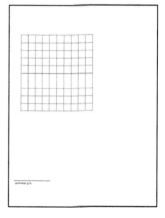

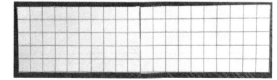

Lesson 9:   Analyze different rectangles and reason about their area.

©2015 Great Minds. eureka math.org
G3-M4-TE-B4-1.3.1-01.2016 -

117

S: 100 square centimeters because I added 50 square centimeters plus 50 square centimeters.
→ 100 square centimeters because that's the total area of the smaller rectangles and that doesn't change when we move them to make the longer rectangle.

T: Let's see if you are right! Answer Problem 2(a). (Allow students time to work.) What multiplication fact can help you find the area of this longer rectangle?

S: 5 × 20.

T: How can you solve this multiplication?

S: We can think of it as 5 times 2 tens. → We could think of it as 5 × (2 × 10), which is the same as (5 × 2) × 10. → We can think of it the same way as before: as 2 equal rectangles.

T: Choose a strategy and use it to answer Problem 2(b). (Allow students time to work.) What is the area of this longer rectangle?

S: 100 square centimeters!

T: Was your prediction about the area of this longer rectangle correct?

S: Yes!

Repeat this process, instructing students to fold 2 columns behind one of the rectangles, so they now have a 5 by 8 rectangle and a 5 by 10 rectangle. They can use their boards to record the total area of the 2 separate rectangles and the area of the longer rectangle that is made by joining the 2 smaller rectangles.

T: What did you notice about the sum of the areas of the 2 small rectangles and the area of the longer rectangle?

S: They're the same!

T: How can we use the areas of 2 small rectangles that form a longer rectangle to find the area of the longer rectangle?

S: Add the areas of the smaller rectangles!

Problem Set (10 minutes)

Students should do their personal best to complete the Problem Set within the allotted 10 minutes. For some classes, it may be appropriate to modify the assignment by specifying which problems they work on first. Some problems do not specify a method for solving. Students should solve these problems using the RDW approach used for Application Problems.

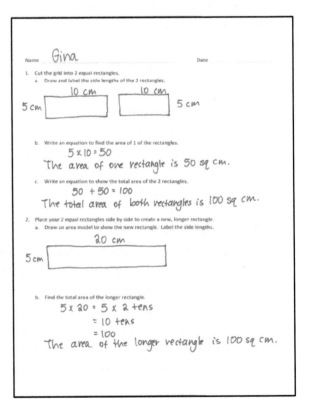

Student Debrief (10 minutes)

Lesson Objective: Analyze different rectangles and reason about their area.

The Student Debrief is intended to invite reflection and active processing of the total lesson experience.

EUREKA
MATH™

Invite students to review their solutions for the Problem Set. They should check work by comparing answers with a partner before going over answers as a class. Look for misconceptions or misunderstandings that can be addressed in the Debrief. Guide students in a conversation to debrief the Problem Set and process the lesson.

Any combination of the questions below may be used to lead the discussion.

- Talk to a partner: In Problem 1(a), how does knowing the side lengths of the grid help you find the side lengths of the small rectangles without counting?

- Did anyone use the break apart and distribute strategy to solve Problem 2(b)? Explain what you broke apart. Why did you make that choice? (Ahead of Lesson 10, which uses the distributive property, ask students how the paper rectangles show the distributive property.)

- Compare the equations you used to solve Problems 1(b) and 2(b). How are they the same? How are they different?

- Explain to a partner how you found the **length** and **width** for the new rectangle in Problem 3(b). If you labeled the width 13 and length 4, how would that change your drawing? How would that affect the area of the rectangle?

- Did anyone multiply the side lengths to solve Problem 3(c)? What strategy did you use to multiply 4 × 13?

- How was Problem 4 different from the other problems?

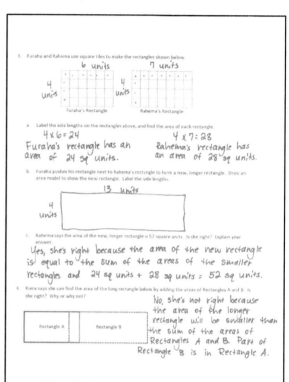

Exit Ticket (3 minutes)

After the Student Debrief, instruct students to complete the Exit Ticket. A review of their work will help with assessing students' understanding of the concepts that were presented in today's lesson and planning more effectively for future lessons. The questions may be read aloud to the students.

Lesson 9:   Analyze different rectangles and reason about their area.

119

©2015 Great Minds. eureka math.org
G3-M4-TE-B4-1.3.1-01.2016 -

Name _____ Date _____

1. Cut the grid into 2 equal rectangles.

   a. Draw and label the side lengths of the 2 rectangles.

   b. Write an equation to find the area of 1 of the rectangles.

   c. Write an equation to show the total area of the 2 rectangles.

2. Place your 2 equal rectangles side by side to create a new, longer rectangle.

   a. Draw an area model to show the new rectangle. Label the side lengths.

   b. Find the total area of the longer rectangle.

Lesson 9:   Analyze different rectangles and reason about their area.

EUREKA
MATH™

3. Furaha and Rahema use square tiles to make the rectangles shown below.

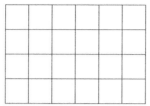

Furaha's Rectangle

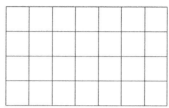

Rahema's Rectangle

a. Label the side lengths on the rectangles above, and find the area of each rectangle.

b. Furaha pushes his rectangle next to Rahema's rectangle to form a new, longer rectangle. Draw an area model to show the new rectangle. Label the side lengths.

c. Rahema says the area of the new, longer rectangle is 52 square units. Is she right? Explain your answer.

4. Kiera says she can find the area of the long rectangle below by adding the areas of Rectangles A and B. Is she right? Why or why not?

Rectangle A

Rectangle B

EUREKA
MATH™

Lesson 9:   Analyze different rectangles and reason about their area.

121

©2015 Great Minds. eureka math.org
G3-M4-TE-B4-1.3.1-01.2016 -

Name _____   Date _____

Lamar uses square tiles to make the 2 rectangles shown below.

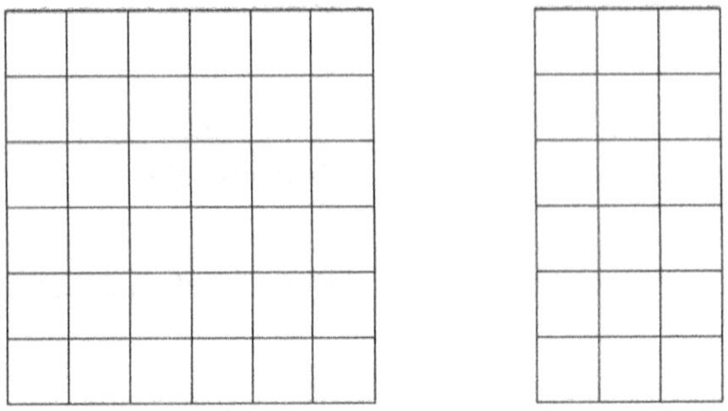

Rectangle A   Rectangle B

1. Label the side lengths of the 2 rectangles.

2. Write equations to find the areas of the rectangles.

Area of Rectangle A: _____   Area of Rectangle B: _____

3. Lamar pushes Rectangle A next to Rectangle B to make a bigger rectangle. What is the area of the bigger rectangle? How do you know?

Lesson 9:   Analyze different rectangles and reason about their area.

EUREKA
MATH™

Name _____   Date _____

1. Use the grid to answer the questions below.

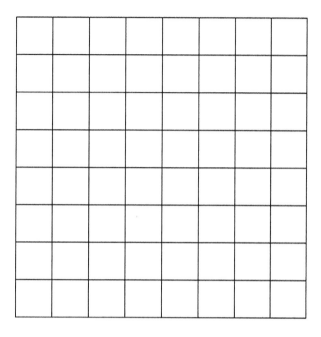

a. Draw a line to divide the grid into 2 equal rectangles. Shade in 1 of the rectangles that you created.

b. Label the side lengths of each rectangle.

c. Write an equation to show the total area of the 2 rectangles.

2. Alexa cuts out the 2 equal rectangles from Problem 1(a) and puts the two shorter sides together.

   a. Draw Alexa's new rectangle and label the side lengths below.

   b. Find the total area of the new, longer rectangle.

   c. Is the area of the new, longer rectangle equal to the total area in Problem 1(c)?
      Explain why or why not.

Lesson 9:   Analyze different rectangles and reason about their area.

EUREKA
MATH

©2015 Great Minds. eureka-math.org
G3-M4-TE-B4-1.3.1-01.2016

small centimeter grid

Lesson 9:   Analyze different rectangles and reason about their area.

125

©2015 Great Minds. eureka math.org
G3-M4-TE-B4-1.3.1-01.2016 -

Lesson 10

Objective: Apply the distributive property as a strategy to find the total area of a large rectangle by adding two products.

Suggested Lesson Structure

■ Fluency Practice   (8 minutes)
▨ Application Problem   (5 minutes)
▢ Concept Development   (37 minutes)
■ Student Debrief   (10 minutes)

   Total Time   **(60 minutes)**

Fluency Practice (8 minutes)

▪ Group Counting **3.OA.1**   (3 minutes)
▪ Find the Unknown Factor **3.OA.4**   (5 minutes)

Group Counting (3 minutes)

Note: Group counting reviews interpreting multiplication as repeated addition.

Instruct students to count forward and backward, occasionally changing the direction of the count.

   ▪ Sixes to 60
   ▪ Sevens to 70
   ▪ Eights to 80
   ▪ Nines to 90

Find the Unknown Factor (5 minutes)

Materials: (S) Personal white board

Note: This fluency activity anticipates finding all possible side lengths of rectangles with areas of 12, 24, 36, 48, and 72 square units in Lesson 11.

   T: (Write $4 \times$ ___ $= 12$.) Find the unknown factor, and say the equation.
   S: $4 \times 3 = 12$.

Continue with the following possible sequence: $6 \times$ ___ $= 12$, $2 \times$ ___ $= 12$, and $3 \times$ ___ $= 12$.

   T: (Write $8 \times$ ___ $= 24$.) Copy my equation on your personal white board, and fill in the unknown factor.
   S: (Write $8 \times 3 = 24$.)

Lesson 10:   Apply the distributive property as a strategy to find the total area of a
large rectangle by adding two products.

©2015 Great Minds. eureka-math.org
G3-M4-TE-B4-1.3.1-01.2016

**EUREKA
MATH™**

Continue with the following possible sequence:

6 × ___ = 24	3 × ___ = 24	6 × ___ = 36	9 × ___ = 36
4 × ___ = 24	4 × ___ = 36	8 × ___ = 72	8 × ___ = 48
9 × ___ = 72	6 × ___ = 48	2 × ___ = 24	12 × ___ = 24
12 × ___ = 36	12 × ___ = 48	12 × ___ = 72	3 × ___ = 36
4 × ___ = 48	6 × ___ = 72	3 × ___ = 72	

**A NOTE ON
12 AS A FACTOR:**

The suggested sequence for this fluency activity helps students solve number sentences with 12 as a factor. While some students might be fluent with these facts, others might rely on the distributive property to write true equations. The expectation is for students to become familiar with 12 as a factor since these equations will be seen in Lesson 11.

Application Problem (5 minutes)

Sonya folds a 6-inch by 6-inch piece of paper into 4 equal parts (shown below). What is the area of 1 of the parts?

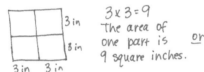

$3 \times 3 = 9$
The area of
one part is   or
9 square inches.

$6 \times 6 = 36$
The piece of paper
has an area of
36 square inches

$36 \div 4 = 9$
The area of
1 folded part is
9 square inches.

Note: This problem reviews the concept of finding area.

Concept Development (37 minutes)

Materials: (S) Personal white board, square centimeter tiles, tiling (Template)

Students start with the tiling template (the partially shaded rectangle shown on the right) in their personal white boards.

T: (Project the tiling template.) There are 3 rectangles we will focus on: the large rectangle (trace its boundary line with your finger), the shaded rectangle (trace its boundary with your finger), and the unshaded rectangle (trace it).

T: Use square centimeter tiles to find the area of the large rectangle. (Allow students time to work.) What is the area of the large rectangle?

S: 48 square centimeters!

T: Use square centimeter tiles to find the side lengths of the shaded rectangle. (Allow students time to work.) What are the side lengths?

S: 5 centimeters and 6 centimeters!

Tiling Template with Sample Work

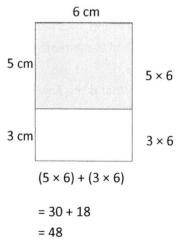

$(5 \times 6) + (3 \times 6)$

$= 30 + 18$

$= 48$

**EUREKA
MATH**™

Lesson 10:   Apply the distributive property as a strategy to find the total area of a
large rectangle by adding two products.

127

©2015 Great Minds. eureka math.org
G3-M4-TE-B4-1.3.1-01.2016 -

T: Label the side lengths. (Allow students time to label the side lengths.) What multiplication expression can you use to find the area of the shaded rectangle?

S: 5 × 6.

T: Write that expression next to the shaded rectangle. (Allow students time to write the expression.) What side length do we already know for the unshaded rectangle?

S: 6 centimeters!

T: Use square centimeter tiles to find the other side length of the unshaded rectangle. (Allow students time to work.) What is the other side length?

S: 3 centimeters!

T: Label the side length. (Allow students time to label the side length.) What multiplication expression can you use to find the area of the unshaded rectangle?

S: 3 × 6.

T: Write that expression next to the unshaded rectangle. (Allow students time to write the expression.) How can we use these two expressions to help us find the area of the large rectangle?

S: We can add them! → The area of the shaded rectangle plus the area of the unshaded rectangle equals the area of the large rectangle.

T: Write a number sentence on your board to show this.

S: (Write (5 × 6) + (3 × 6).)

T: Read your number sentence to a partner, and then find its value. (Allow students time to solve.) What is the area of the large rectangle?

S: 48 square centimeters!

T: Is that the answer you got when you **tiled** the large rectangle?

S: Yes!

T: Write the value of the length of the large rectangle as an addition expression.

MP.7

S: (Write 5 + 3.)

T: What will you multiply by to find the area?

S: 6.

T: Write that in your expression. Where should we put parentheses?

S: Around 5 + 3 because we need to add first to find the side length; then, we can multiply.

T: Add the parentheses to your expression. What is 5 + 3?

S: 8.

T: What is the new expression?

S: 8 × 6.

T: What is the area?

S: 48 square centimeters!

T: Is that the same answer we just found?

S: Yes!

Distributive Property Expressions

8 × 6
(5 + 3) × 6
(5 × 6) + (3 × 6)

Lesson 10: Apply the distributive property as a strategy to find the total area of a large rectangle by adding two products.

©2015 Great Minds. eureka-math.org
G3-M4-TE-B4-1.3.1-01.2016

EUREKA MATH

T: (Record distributive property expressions as shown on previous page.) How are these three expressions related?

S: They all show the area of the large rectangle. → Oh look, they show the break apart and distribute strategy! → Yeah, they show that the side length 8 is broken apart into 5 plus 3. Then, 5 and 3 are multiplied by the other side length, 6.

MP.7

T: Discuss with a partner how the large rectangle on your personal white board also shows the break apart and distribute strategy.

S: (Discuss.)

Repeat the process with the following possible suggestions, providing pictures of rectangles with grid lines:

- A 15 by 8 rectangle. (Students can partition as $(10 + 5) \times 8$. This helps students see that this strategy is helpful when they cannot multiply the side lengths because they do not know these facts.)

- An 18 by 9 rectangle. (Students can decompose as double 9×9 or $(10 + 8) \times 9$.)

T: We broke apart the 18 by 9 rectangle into two 9 by 9 rectangles. What other ways could we break apart this rectangle?

S: I would do 10 by 9 and 8 by 9 rectangles.

T: Explain to a partner the process you use to decide how to break apart a side length.

S: I look for facts I know. → I try to find a way to make a 5 or 10 because they're easy facts.

Problem Set (10 minutes)

Students should do their personal best to complete the Problem Set within the allotted 10 minutes. For some classes, it may be appropriate to modify the assignment by specifying which problems they work on first. Some problems do not specify a method for solving. Students should solve these problems using the RDW approach used for Application Problems.

Student Debrief (10 minutes)

Lesson Objective: Apply the distributive property as a strategy to find the total area of a large rectangle by adding two products.

The Student Debrief is intended to invite reflection and active processing of the total lesson experience.

> **NOTES ON MULTIPLE MEANS OF ACTION AND EXPRESSION:**
>
> Consider directing students who may not complete the Problem Set within the allotted time to Problem 2 for valuable application and demonstration of understanding of today's objective. Offer planning and strategy development support to learners, if needed. Model a think-aloud with two or more possibilities, reason about the selection, and solve.

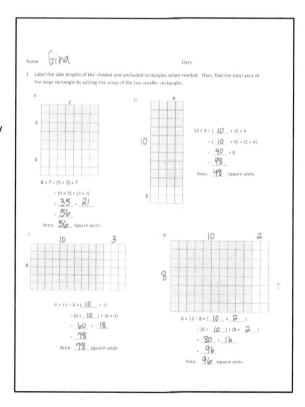

EUREKA MATH™

Lesson 10:   Apply the distributive property as a strategy to find the total area of a large rectangle by adding two products.

129

©2015 Great Minds. eureka math.org
G3-M4-TE-B4-1.3.1-01.2016

Invite students to review their solutions for the Problem Set. They should check work by comparing answers with a partner before going over answers as a class. Look for misconceptions or misunderstandings that can be addressed in the Debrief. Guide students in a conversation to debrief the Problem Set and process the lesson.

Any combination of the questions below may be used to lead the discussion.

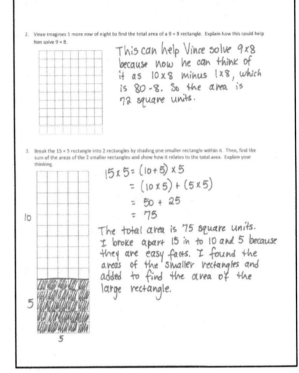

- How is the rectangle in Problem 1(a) similar to the rectangle you **tiled** in today's lesson? How is it different?

- What are the side lengths of the large rectangle in Problem 1(c)? Can you multiply these side lengths to find the area? How does the break apart and distribute strategy help you?

- Without multiplying the side lengths of the large rectangle in Problem 1(d), how could you check to make sure your answer is right? (Students might say count the squares or skip-count by eight 12 times.) Discuss with a partner which strategy is most efficient—either counting squares, skip-counting, or using the break apart and distribute strategy.

- How was setting up and solving Problem 2 different from the other problems?

- What side length did you break apart in Problem 3, and how did you break it apart? Why?

- With a partner, list as many possibilities as you can for how you could use the break apart and distribute strategy to find the area of a rectangle with side lengths of 20 and 7. Can we break it into 3 parts if we want to? Which one of your possibilities would you use to solve this problem? Why?

Exit Ticket (3 minutes)

After the Student Debrief, instruct students to complete the Exit Ticket. A review of their work will help with assessing students' understanding of the concepts that were presented in today's lesson and planning more effectively for future lessons. The questions may be read aloud to the students.

Lesson 10: Apply the distributive property as a strategy to find the total area of a large rectangle by adding two products.

©2015 Great Minds. eureka-math.org
G3-M4-TE-B4-1.3.1-01.2016

EUREKA MATH™

Name _____ Date _____

1. Label the side lengths of the shaded and unshaded rectangles when needed. Then, find the total area of the large rectangle by adding the areas of the two smaller rectangles.

a.

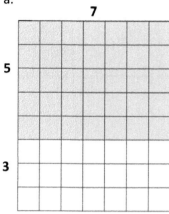

7

5

3

$8 \times 7 = (5 + 3) \times 7$

$= (5 \times 7) + (3 \times 7)$

$= \underline{\hspace{1cm}} + \underline{\hspace{1cm}}$

$= \underline{\hspace{1cm}}$

Area: _____ square units

b.

4

2

$12 \times 4 = (\underline{\hspace{1cm}} + 2) \times 4$

$= (\underline{\hspace{1cm}} \times 4) + (2 \times 4)$

$= \underline{\hspace{1cm}} + 8$

$= \underline{\hspace{1cm}}$

Area: _____ square units

c.

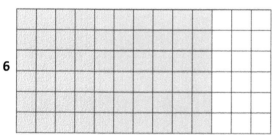

6

$6 \times 13 = 6 \times (\underline{\hspace{1cm}} + 3)$

$= (6 \times \underline{\hspace{1cm}}) + (6 \times 3)$

$= \underline{\hspace{1cm}} + \underline{\hspace{1cm}}$

$= \underline{\hspace{1cm}}$

Area: _____ square units

d.

$8 \times 12 = 8 \times (\underline{\hspace{1cm}} + \underline{\hspace{1cm}})$

$= (8 \times \underline{\hspace{1cm}}) + (8 \times \underline{\hspace{1cm}})$

$= \underline{\hspace{1cm}} + \underline{\hspace{1cm}}$

$= \underline{\hspace{1cm}}$

Area: _____ square units

Lesson 10: Apply the distributive property as a strategy to find the total area of a large rectangle by adding two products.

131

©2015 Great Minds. eureka math.org
G3-M4-TE-B4-1.3.1-01.2016 -

2. Vince imagines 1 more row of eight to find the total area of a 9 × 8 rectangle. Explain how this could help him solve 9 × 8.

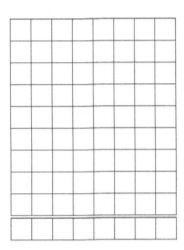

3. Break the 15 × 5 rectangle into 2 rectangles by shading one smaller rectangle within it. Then, find the sum of the areas of the 2 smaller rectangles and show how it relates to the total area. Explain your thinking.

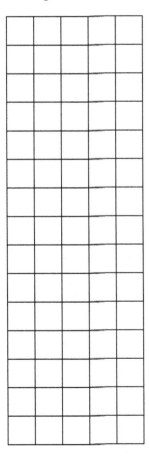

Lesson 10:   Apply the distributive property as a strategy to find the total area of a large rectangle by adding two products.

©2015 Great Minds. eureka-math.org
G3-M4-TE-B4-1.3.1-01.2016

EUREKA
MATH™

Name _____   Date _____

Label the side lengths of the shaded and unshaded rectangles. Then, find the total area of the large rectangle by adding the areas of the 2 smaller rectangles.

1.

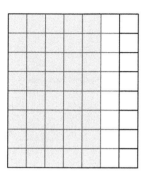

$8 \times 7 = 8 \times ($ _____ + _____ $)$

$= (8 \times$ _____ $) + (8 \times$ _____ $)$

$=$ _____ + _____

$=$ _____

Area: _____ square units

2.

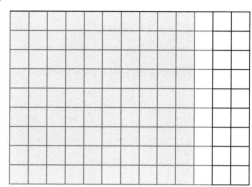

$9 \times 13 = 9 \times ($ _____ + _____ $)$

$= ($ _____ $\times$ _____ $) + ($ _____ $\times$ _____ $)$

$=$ _____ + _____

$=$ _____

Area: _____ square units

EUREKA MATH™   **Lesson 10:**   Apply the distributive property as a strategy to find the total area of a large rectangle by adding two products.   **133**

©2015 Great Minds. eureka math.org
G3-M4-TE-B4-1.3.1-01.2016   -

Name _____   Date _____

1. Label the side lengths of the shaded and unshaded rectangles. Then, find the total area of the large rectangle by adding the areas of the 2 smaller rectangles.

a.

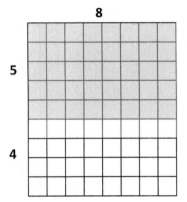

$9 \times 8 = (5 + 4) \times 8$

$= (5 \times 8) + (4 \times 8)$

$= _____ + _____$

$= _____$

Area: _____ square units

b.

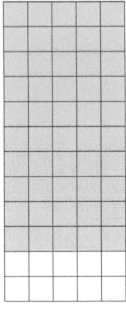

$12 \times 5 = (_____ + 2) \times 5$

$= (_____ \times 5) + (2 \times 5)$

$= _____ + 10$

$= _____$

Area: _____ square units

c.

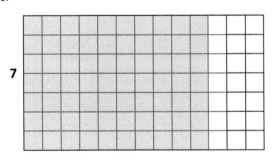

$7 \times 13 = 7 \times (_____ + 3)$

$= (7 \times _____) + (7 \times 3)$

$= _____ + _____$

$= _____$

Area: _____ square units

d.

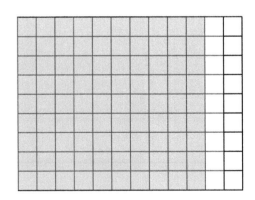

$9 \times 12 = 9 \times (_____ + _____)$

$= (9 \times _____) + (9 \times _____)$

$= _____ + _____$

$= _____$

Area: _____ square units

Lesson 10:   Apply the distributive property as a strategy to find the total area of a large rectangle by adding two products.

EUREKA MATH™

2. Finn imagines 1 more row of nine to find the total area of 9 × 9 rectangle. Explain how this could help him solve 9 × 9.

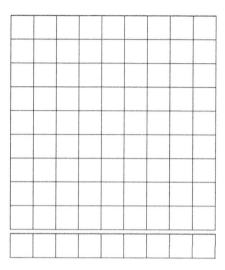

3. Shade an area to break the 16 × 4 rectangle into 2 smaller rectangles. Then, find the sum of the areas of the 2 smaller rectangles to find the total area. Explain your thinking.

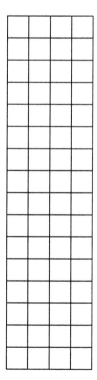

EUREKA
MATH

Lesson 10:   Apply the distributive property as a strategy to find the total area of a
             large rectangle by adding two products.

©2015 Great Minds. eureka math.org
G3-M4-TE-B4-1.3.1-01.2016 -

135

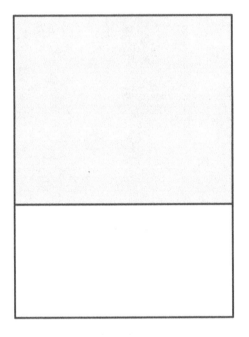

tiling

Lesson 10: Apply the distributive property as a strategy to find the total area of a large rectangle by adding two products.

Lesson 11

Objective: Demonstrate the possible whole number side lengths of rectangles with areas of 24, 36, 48, or 72 square units using the associative property.

Suggested Lesson Structure

- ■ Fluency Practice   (13 minutes)
- ▨ Application Problem   (5 minutes)
- ▢ Concept Development   (32 minutes)
- ■ Student Debrief   (10 minutes)

  Total Time   **(60 minutes)**

Fluency Practice (13 minutes)

- Group Counting **3.OA.1**   (3 minutes)
- Find the Unknown Factor **3.OA.4**   (5 minutes)
- Find the Area **3.MD.7**   (5 minutes)

Group Counting (3 minutes)

Note: Group counting reviews interpreting multiplication as repeated addition.

Instruct students to count forward and backward, occasionally changing the direction of the count.

- ▪ Sixes to 60
- ▪ Sevens to 70
- ▪ Eights to 80
- ▪ Nines to 90

Find the Unknown Factor (5 minutes)

Materials: (S) Personal white board

Note: This fluency activity anticipates the objective of today's lesson.

  T: (Write $6 \times$ ___ $= 12$.) Find the unknown factor, and say the equation.
  S: $6 \times 2 = 12$.

Continue with the following possible sequence: $4 \times$ ___ $= 12$, $2 \times$ ___ $= 12$, and $3 \times$ ___ $= 12$.

Lesson 11:   Demonstrate the possible whole number side lengths of rectangles
              with areas of 24, 36, 48, or 72 square units using the associative
              property.

©2015 Great Minds. eureka math.org
G3-M4-TE-B4-1.3.1-01.2016 -

T: (Write 3 × ___ = 24.) Copy my equation on your personal white board, and fill in the unknown factor.

S: (Write 3 × 8 = 24.)

Continue with the following possible sequence: 4 × ___ = 24, 8 × ___ = 24, 6 × ___ = 36, 4 × ___ = 36, 6 × ___ = 24, 9 × ___ = 36, 9 × ___ = 72, 6 × ___ = 48, 8 × ___ = 72, 8 × ___ = 48, and 2 × ___ = 24.

Find the Area (5 minutes)

Materials: (S) Personal white board

Note: This fluency activity reviews using the distributive property from Lesson 10.

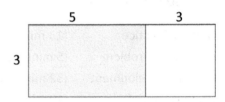

T: (Project the rectangle as shown.) On your personal white board, write an expression that we could use to find the area of the shaded rectangle.

S: (Write 3 × 5.)

T: On your board, write an expression that we could use to find the area of the unshaded rectangle.

S: (Write 3 × 3.)

T: How can you use these expressions to find the area of the large rectangle?

S: Add them!

T: Write an equation showing the sum of the shaded and unshaded rectangles. Below it, write the area of the entire rectangle.

S: (Write 15 + 9 = 24 and Area: 24 square units.)

Sample Work

$(3 × 5) + (3 × 3)$

$= 15 + 9$

$= 24$

Continue with the following possible sequence:
9 × 5 = (5 × 5) + (4 × 5), 13 × 4 = (10 × 4) + (3 × 4), and 17 × 3 = (10 × 3) + (7 × 3).

NOTES ON MULTIPLE MEANS OF ENGAGEMENT:

Alternatively, challenge students working above grade level with this *length unknown* version:

One fourth of the banquet table has an area of 9 square feet. If the width of the table is 3 feet, what is the length? What is the area of the table?

Application Problem (5 minutes)

The banquet table in a restaurant measures 3 feet by 6 feet. For a large party, workers at the restaurant place 2 banquet tables side by side to create 1 long table. Find the area of the new, longer table.

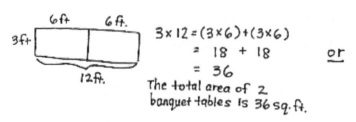

or

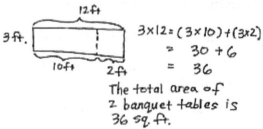

Lesson 11: Demonstrate the possible whole number side lengths of rectangles with areas of 24, 36, 48, or 72 square units using the associative property.

©2015 Great Minds. eureka-math.org
G3-M4-TE-B4-1.3.1-01.2016

EUREKA MATH

Note: This problem reviews Lesson 10's concept of applying the distributive property to find the total area of a large rectangle by adding two products. It also reviews factors of 36 and multiples of 12 that lead into the Concept Development.

Concept Development (32 minutes)

Materials: (S) Personal white board

12

3

T: Write an expression to show how to find the area of a rectangle with side lengths 3 and 12.

S: (Write 3 × 12.)

T: In the Application Problem, you found that 3 times 12 is…?

S: 36.

T: So, the area of this rectangle is…?

S: 36 square units!

T: (Write 3 × (2 × 6).) Why is this expression equal to the one you just wrote?

S: Because you just wrote 12 as 2 × 6.

T: Write this expression on your personal white board with the parentheses in a different place. At my signal, show me your board. (Signal.)

S: (Show (3 × 2) × 6.)

T: Solve 3 × 2, and write the new expression on your board. (Allow students time to work.) Whisper the new expression to a partner.

S: 6 × 6.

T: What new side lengths did we find for a rectangle with an area of 36 square units?

S: 6 and 6.

T: Let's look at our expression, (3 × 2) × 6, again. Use the commutative property and switch the order of the factors in the parentheses.

S: (Write (2 × 3) × 6.)

T: Will you be able to find new side lengths by moving the parentheses?

S: (Write 2 × (3 × 6).) Yes, they will be 2 and 18.

T: (Write 3 × (3 × 4).) Why is this expression equal to our first one, 3 × 12?

S: Because you wrote 12 as 3 × 4.

T: Write this expression on your board with the parentheses in a different place. At my signal, show me your board. (Signal.)

S: (Show (3 × 3) × 4.)

T: Solve 3 × 3 and write the expression on your board. (Allow students time to work.) Whisper the new expression to a partner.

S: 9 × 4.

T: What new side lengths did we find for a rectangle with an area of 36 square units?

S: 9 and 4.

T: Let's look at our expression (3 × 3) × 4 again. If I use the commutative property and switch the order of the factors in the parentheses, will I be able to find new side lengths by moving the parentheses?

S: No, it will still be 9 and 4. → No, because both factors in the parentheses are 3, so switching their order won't change the numbers you get when you move the parentheses.

T: Do you think we found all the possible **whole number** side lengths for this rectangle?

S: Yes. → I'm not sure.

T: Let's look at our side lengths. Do you have a side length of 1?

S: No! We forgot the easiest one. → It's 1 and 36.

T: Do we have a side length of 2?

S: Yes.

T: 3?

S: Yes.

T: Work with a partner to look at the rest of your side lengths to see if you have the numbers 4 through 10. (Allow students time to work.) Which of these numbers, 4 through 10, aren't included in your side lengths?

MP.3

S: 5, 7, 8, and 10.

T: Discuss with a partner why these numbers aren't in your list of side lengths.

S: 5, 7, 8, and 10 can't be side lengths because there aren't any whole numbers we can multiply these numbers by to get 36.

T: Would any two-digit number times two-digit number work?

S: No, they would be too big. → No, because we know 10 × 10 equals 100, and that's greater than 36.

T: Now, do you think we found all the possible side whole number side lengths for a rectangle with an area of 36 square units?

S: Yes!

Repeat the process with rectangles that have areas of 24, 48, and 72 square units.

Problem Set (10 minutes)

Students should do their personal best to complete the Problem Set within the allotted 10 minutes. For some classes, it may be appropriate to modify the assignment by specifying which problems they work on first. Some problems do not specify a method for solving. Students should solve these problems using the RDW approach used for Application Problems.

NOTES ON MULTIPLE MEANS OF REPRESENTATION:

Extend Problem 1 for students working above grade level by inviting experimentation and choice in placing parentheses, as well as number order, in the multiplication sentences. For example, ask, "What would happen if we changed it to 4 × 6 × 2?" Encourage students to discuss or journal about their discoveries.

Assist English language learners by rephrasing Problem 4 in multiple ways. Ask the following: "How does the difference between the length and width of the rectangle change the shape?"

Lesson 11:   Demonstrate the possible whole number side lengths of rectangles with areas of 24, 36, 48, or 72 square units using the associative property.
©2015 Great Minds. eureka-math.org
G3-M4-TE-B4-1.3.1-01.2016

Student Debrief (10 minutes)

Lesson Objective: Demonstrate the possible whole number side lengths of rectangles with areas of 24, 36, 48, or 72 square units using the associative property.

The Student Debrief is intended to invite reflection and active processing of the total lesson experience.

Invite students to review their solutions for the Problem Set. They should check work by comparing answers with a partner before going over answers as a class. Look for misconceptions or misunderstandings that can be addressed in the Debrief. Guide students in a conversation to debrief the Problem Set and process the lesson.

Any combination of the questions below may be used to lead the discussion.

- Turn your paper horizontally and look at Problem 1. What property does this show?

- Share your answer to Problem 2 with a partner.

- Discuss your answer to Problem 4 with a partner. What would the rectangle look like if the difference between side lengths was 0? How do you know?

- Compare your answer to Problem 4(c) with a partner's. Did you both come up with the same side lengths? Why or why not?

- Explain to a partner how to use the strategy we learned today to find all possible **whole number** side lengths for a rectangle with an area of 60 square units.

Exit Ticket (3 minutes)

After the Student Debrief, instruct students to complete the Exit Ticket. A review of their work will help you assess the students' understanding of the concepts that were presented in the lesson today and plan more effectively for future lessons. You may read the questions aloud to the students.

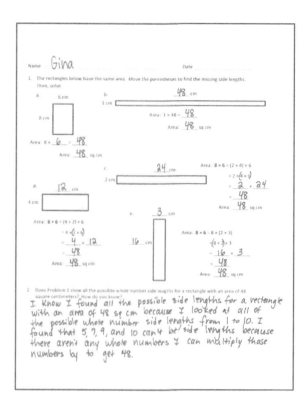

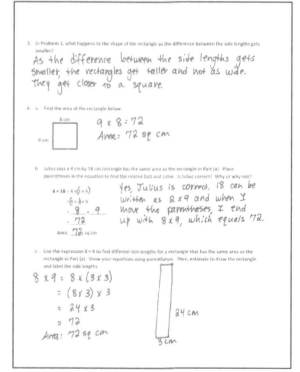

Lesson 11:   Demonstrate the possible whole number side lengths of rectangles with areas of 24, 36, 48, or 72 square units using the associative property.

©2015 Great Minds. eureka math.org
G3-M4-TE-B4-1.3.1-01.2016 -

141

Name _____   Date _____

1. The rectangles below have the same area. Move the parentheses to find the unknown side lengths. Then, solve.

a.

6 cm

8 cm

Area: 8 × _____ = _____

Area: _____ sq cm

b.

_____ cm

1 cm

Area: 1 × 48 = _____

Area: _____ sq cm

Area: **8 × 6** = (2 × 4) × 6

= 2 × 4 × 6

= _____ × _____

= _____

Area: _____ sq cm

c.

_____ cm

2 cm

d.

_____ cm

4 cm

Area: **8 × 6** = (4 × 2) × 6

= 4 × 2 × 6

= _____ × _____

= _____

Area: _____ sq cm

e.

_____ cm

_____ cm

Area: **8 × 6** = 8 × (2 × 3)

= 8 × 2 × 3

= _____ × _____

= _____

Area: _____ sq cm

2. Does Problem 1 show all the possible whole number side lengths for a rectangle with an area of 48 square centimeters? How do you know?

Lesson 11:   Demonstrate the possible whole number side lengths of rectangles with areas of 24, 36, 48, or 72 square units using the associative property.
©2015 Great Minds. eureka-math.org
G3-M4-TE-B4-1.3.1-01.2016

EUREKA
MATH

3. In Problem 1, what happens to the shape of the rectangle as the difference between the side lengths gets smaller?

4. a. Find the area of the rectangle below.

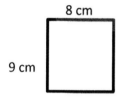

8 cm

9 cm

b. Julius says a 4 cm by 18 cm rectangle has the same area as the rectangle in Part (a). Place parentheses in the equation to find the related fact and solve. Is Julius correct? Why or why not?

$4 \times 18 = 4 \times 2 \times 9$

$= 4 \times 2 \times 9$

$= \underline{\hspace{1cm}} \times \underline{\hspace{1cm}}$

$= \underline{\hspace{1cm}}$

Area: _____ sq cm

c. Use the expression 8 × 9 to find different side lengths for a rectangle that has the same area as the rectangle in Part (a). Show your equations using parentheses. Then, estimate to draw the rectangle and label the side lengths.

EUREKA
MATH™

Lesson 11: Demonstrate the possible whole number side lengths of rectangles with areas of 24, 36, 48, or 72 square units using the associative property.

©2015 Great Minds. eureka math.org
G3-M4-TE-B4-1.3.1-01.2016 -

143

Name _____ Date _____

1. Find the area of the rectangle.

8 cm

8 cm

2. The rectangle below has the same area as the rectangle in Problem 1. Move the parentheses to find the unknown side lengths. Then, solve.

_____ cm

_____ cm

Area: **8 × 8** = (4 × 2) × 8

= 4 × 2 × 8

= _____ × _____

= _____

Area: _____ sq cm

Lesson 11: Demonstrate the possible whole number side lengths of rectangles with areas of 24, 36, 48, or 72 square units using the associative property.

©2015 Great Minds. eureka-math.org
G3-M4-TE-B4-1.3.1-01.2016

Name _____ Date _____

1. The rectangles below have the same area. Move the parentheses to find the unknown side lengths. Then, solve.

36 cm

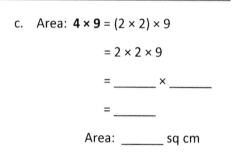

1 cm

b. Area: $1 \times 36 =$ _____

Area: _____ sq cm

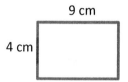

9 cm

4 cm

a. Area: $4 \times$ _____ $=$ _____

Area: _____ sq cm

_____ cm

2 cm

c. Area: $\mathbf{4 \times 9} = (2 \times 2) \times 9$

$= 2 \times 2 \times 9$

$=$ _____ $\times$ _____

$=$ _____

Area: _____ sq cm

_____ cm

_____ cm

d. Area: $\mathbf{4 \times 9} = 4 \times (3 \times 3)$

$= 4 \times 3 \times 3$

$=$ _____ $\times$ _____

$=$ _____

Area: _____ sq cm

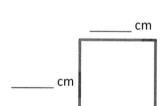

e. Area: $\mathbf{12 \times 3} = (6 \times 2) \times 3$

$= 6 \times 2 \times 3$

_____ cm

$=$ _____ $\times$ _____

_____ cm

$=$ _____

Area: _____ sq cm

2. Does Problem 1 show all the possible whole number side lengths for a rectangle with an area of 36 square centimeters? How do you know?

EUREKA
MATH™

Lesson 11:   Demonstrate the possible whole number side lengths of rectangles
with areas of 24, 36, 48, or 72 square units using the associative
property.
©2015 Great Minds. eureka-math.org
G3-M4-TE-B4-1.3.1-01.2016 -

145

3. a. Find the area of the rectangle below.

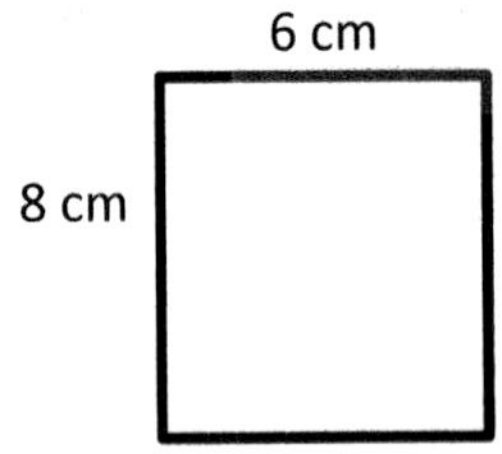

6 cm

8 cm

b. Hilda says a 4 cm by 12 cm rectangle has the same area as the rectangle in Part (a). Place parentheses in the equation to find the related fact and solve. Is Hilda correct? Why or why not?

4 × 12 = 4 × 2 × 6

= 4 × 2 × 6

= _____ × _____

= _____

Area: _____ sq cm

c. Use the expression 8 × 6 to find different side lengths for a rectangle that has the same area as the rectangle in Part (a). Show your equations using parentheses. Then, estimate to draw the rectangle and label the side lengths.

Lesson 11: Demonstrate the possible whole number side lengths of rectangles with areas of 24, 36, 48, or 72 square units using the associative property.
©2015 Great Minds. eureka-math.org
G3-M4-TE-B4-1.3.1-01.2016

EUREKA MATH™

3
GRADE

Mathematics Curriculum

Topic D

Applications of Area Using Side Lengths of Figures

3.MD.6, 3.MD.7, 3.MD.5

Focus Standards:	3.MD.6	Measure areas by counting unit squares (square cm, square m, square in, square ft, and improvised units).
	3.MD.7	Relate area to the operations of multiplication and addition. a. Find the area of a rectangle with whole-number side lengths by tiling it, and show that the area is the same as would be found by multiplying the side lengths. b. Multiply side lengths to find areas of rectangles with whole-number side lengths in the context of solving real world and mathematical problems, and represent whole-number products as rectangular areas in mathematical reasoning. c. Use tiling to show in a concrete case that the area of a rectangle with whole-number side lengths a and $b + c$ is the sum of $a \times b$ and $a \times c$. Use area models to represent the distributive property in mathematical reasoning. d. Recognize area as additive. Find areas of rectilinear figures by decomposing them into non-overlapping rectangles and adding the areas of the non-overlapping parts, applying this technique to solve real world problems.
Instructional Days:	5	
Coherence -Links from:	G2–M2	Addition and Subtraction of Length Units
	G3–M1	Properties of Multiplication and Division and Solving Problems with Units of 2–5 and 10
	G3–M3	Multiplication and Division with Units of 0, 1, 6–9, and Multiples of 10
-Links to:	G4–M3	Multi-Digit Multiplication and Division
	G4–M7	Exploring Multiplication

Topic D requires students to synthesize and apply their knowledge of area. Lesson 12 begins the topic with an emphasis on real-world applications by providing students with opportunities to apply their understanding of area to solving word problems. Students may practice *unknown product, group size unknown*, and *number of groups unknown* types of problems. (See examples of problem types in the chart on page 19 of the Geometric Measurement Progression.) The word problems provide a stepping-stone for the real-world, project-based application of area to composite shapes and the area floor plan in this topic.

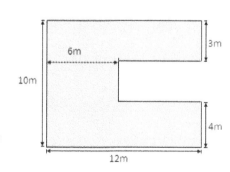

Lessons 13 and 14 introduce students to finding the area of composite shapes. They learn to find the unknown measurements using the given side lengths and then make decisions about whether to decompose the tiled region into smaller rectangles and add the areas (**3.MD.7c**) or complete the composite figures and then subtract.

In Lessons 15 and 16, students apply their work with composite shapes from the previous two lessons to a real-word application, determining areas of rooms in a given floor plan.

A Teaching Sequence Toward Mastery of Applications of Area Using Side Lengths of Figures
Objective 1: **Solve word problems involving area.**   (Lesson 12)
Objective 2: **Find areas by decomposing into rectangles or completing composite figures to form rectangles.**   (Lessons 13–14)
Objective 3: **Apply knowledge of area to determine areas of rooms in a given floor plan.**   (Lessons 15–16)

Lesson 12

Objective: Solve word problems involving area.

Suggested Lesson Structure

■ Fluency Practice   (15 minutes)
▨ Application Problem   (5 minutes)
▢ Concept Development   (30 minutes)
■ Student Debrief   (10 minutes)

   Total Time   **(60 minutes)**

Fluency Practice (15 minutes)

▪ Group Counting **3.OA.1**   (3 minutes)
▪ Multiply by 7 **3.OA.7**   (7 minutes)
▪ Find the Side Length **3.MD.7**   (5 minutes)

Group Counting (3 minutes)

Note: Group counting reviews interpreting multiplication as repeated addition.

Instruct students to count forward and backward, occasionally changing the direction of the count.

   ▪   Fours to 40
   ▪   Sixes to 60
   ▪   Eights to 80
   ▪   Nines to 90

Multiply by 7 (7 minutes)

Materials: (S) Multiply by 7 (6–10) Pattern Sheet

Note: This activity builds fluency with multiplication facts using units of 7. It works toward students knowing from memory all products of two one-digit numbers. See Lesson 2 for the directions for administration of a Multiply-By Pattern Sheet.

   T: (Write $7 \times 7 =$ ___.) Let's skip-count up by sevens. (Count with fingers to 7 as students count.)
   S: 7, 14, 21, 28, 35, 42, 49.
   T: Let's see how we can skip-count down to find the answer, too. (Show 10 fingers.) Start at 70. (Count down with your fingers as students say numbers.)
   S: 70, 63, 56, 49.

Continue with the following possible sequence: 9 × 7, 6 × 7, and 8 × 7.

T: (Distribute Multiply by 7 (6–10) Pattern Sheet.) Let's practice multiplying by 7. Be sure to work left to right across the page.

Find the Side Length (5 minutes)

Materials: (S) Personal white board

Note: This fluency activity reviews the relationship between side lengths and area.

T: (Project a rectangle with a width of 2 units and an unknown length. Inside the rectangle, write *Area = 10 square units*.) Say the area of the rectangle.

S: 10 square units.

T: What's the width of the rectangle?

S: 2 units.

T: (Write 2 units × __ units = 10 square units.) On your personal white board, complete the equation, filling in the unknown length.

S: (Write 2 units × 5 units = 10 square units.)

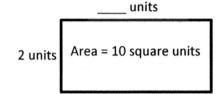

Continue with the possible following sequence: 1 unit × __ units = 8 square units, 5 units × __ units = 15 square units, 3 units × __ units = 18 square units, and 6 units × __ units = 24 square units.

Application Problem (5 minutes)

a. Find the area of a 6 meter by 9 meter rectangle.

b. Use the side lengths, 6 m × 9 m, to find different side lengths for a rectangle that has the same area. Show your equations using parentheses. Then estimate to draw the rectangle and label the side lengths.

a.

9 m

6m

6 × 9 = 54
The area of the rectangle is 54 square meters.

b. 6 × 9 = 6 × (3 × 3)
     = (6 × 3) × 3
     = 18 × 3
     = 54

18 m

3 m

The side lengths of the rectangle are 18 meters and 3 meters.

Note: This problem reviews using the associative property to generate whole number side lengths of rectangles with a given area.

Lesson 12:   Solve word problems involving area.

EUREKA
MATH

Concept Development (30 minutes)

Materials: (S) Personal white board

Problem 1: Solve area word problems with 1 side length unknown.

Write or project the following problem: The area of Theo's banner is 32 square feet. If the length of his banner measures 4 feet, how wide is his banner?

T: What information do we know?

S: The area and length of Theo's banner.

T: What information do we not know?

S: The width.

T: I'll draw an area model and use a letter for what we don't know. (Draw an incorrectly scaled model such as the one shown on the right.)

4 ft

w   Area = 32 sq ft

MP.6

T: If the length is 4 feet and the area is 32 square feet, can the width be less than 4 feet?

S: No, the width needs to be more than 4 feet. → The width should be more than 4 feet because 4 times 4 only equals 16, but the area is 32 square feet.

T: Talk to your partner: Is the area model I drew an accurate representation of the rectangle in the problem? How do you know?

S: No, because the width should be much longer than the length.

T: Work with your partner to correctly redraw my area model on your board.

S: (Draw as shown on the right.)

T: How can we find the value of w?

S: Divide 32 by 4.

T: Write a division equation to find the value of w.

S: (Write $32 \div 4 = w$.)

T: What is the value of w?

S: 8.

T: So, the width of Theo's banner is just 8? 8 what?

S: 8 feet!

4 ft

w   Area = 32 sq ft

Repeat the process with the following suggestions:

- The area of a piece of paper is 72 square inches. Margo measures the length of the paper and says it is 8 inches. What is the width of the piece of paper?

- Jillian needs to draw a rectangle with an area of 56 square centimeters and a width of 7 centimeters. What is the length of the rectangle that Jillian needs to draw?

Problem 2: Choose a strategy to find the area of a larger rectangle.

Write or project the following problem: Amir is getting carpet in his bedroom, which measures 7 feet by 15 feet. How many square feet of carpet will Amir need?

T: Draw an area model to represent Amir's bedroom. Write an expression that shows how to find the area.

S: (Draw as shown to the right.)

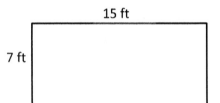

15 ft

7 ft

Area: 7 × 15

T: Talk to your partner: How can we find the area of Amir's bedroom since the measurements are so large?

S: We can break the room up into two smaller rectangles and add their areas together. → We can also break apart one of the factors in 7 × 15 to come up with a multiplication sentence that is easier to solve.

T: Decide with your partner which strategy you'll use to find the area. Then, solve.

S: (Decide on a strategy and solve.)

T: What is the area of Amir's bedroom?

S: 105 square feet!

Invite students to share which strategy they chose and why; ask them to articulate how they used the strategy to solve the problem. For the break apart and distribute strategy, students may have broken apart the rectangle several different ways.

Continue with the following suggested examples, encouraging students to try different strategies:

- Maya helps her family tile the bathroom wall. It measures 12 feet by 11 feet. How many square foot tiles does Maya need to cover the wall?

- Francis washes all of the windows outside his parents' bookstore; there are 5 windows, each one is 6 feet wide and 8 feet high. What is the total area of the windows that Francis washes?

Problem Set (10 minutes)

Students should do their personal best to complete the Problem Set within the allotted 10 minutes. For some classes, it may be appropriate to modify the assignment by specifying which problems they work on first. Some problems do not specify a method for solving. Students should solve these problems using the RDW approach used for Application Problems.

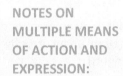

NOTES ON
MULTIPLE MEANS
OF ACTION AND
EXPRESSION:

Offer planning and strategy development support to learners, if needed. Some learners may use a method simply because they are not fluent in an alternative method. Model a think-aloud in which two or more strategies are possible, reason about your selection, and solve. This may take more time than allotted here. Consider pre-teaching in order to preserve the pace of the lesson and to maximize every student's participation.

NOTES ON
MULTIPLE MEANS
OF ENGAGEMENT:

During the Problem Set, extend Problem 4 to students working above grade level. Have students model all possible rectangles with an area of 64. Or, have students model up to eight ways of breaking their rectangle (Part b) into two smaller rectangles. Make it an exciting, perhaps timed, competition. Always offer challenges and extensions to learners as alternatives rather than additional busy work.

EUREKA MATH

Student Debrief (10 minutes)

Lesson Objective: Solve word problems involving area.

The Student Debrief is intended to invite reflection and active processing of the total lesson experience.

Invite students to review their solutions for the Problem Set. They should check work by comparing answers with a partner before going over answers as a class. Look for misconceptions or misunderstandings that can be addressed in the Debrief. Guide students in a conversation to debrief the Problem Set and process the lesson.

Any combination of the questions below may be used to lead the discussion.

- What shape is the sticky note in Problem 1? How do you know?

- Share student explanations to Problem 2(b).

- What is another way the artist's mural in Problem 3 could have been broken apart?

- How did you identify Alana's pattern in Problem 4?

- Discuss how you found the area of two pieces of Jermaine's paper in Problem 5. Why was it necessary to find the unknown side length first? Are there any other ways to find the area of the two pieces of paper? (81 − 27 = 54)

- How were all of today's word problems related? Does the unknown in a problem change the way you solve it? Why or why not?

Exit Ticket (3 minutes)

After the Student Debrief, instruct students to complete the Exit Ticket. A review of their work will help with assessing students' understanding of the concepts that were presented in today's lesson and planning more effectively for future lessons. The questions may be read aloud to the students.

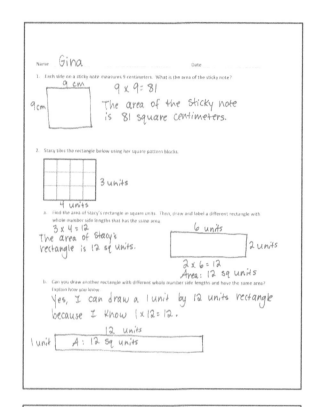

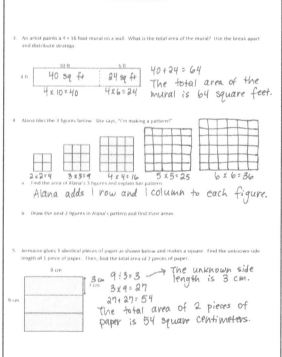

Multiply.

$7 \times 1 = $ _____ $7 \times 2 = $ _____ $7 \times 3 = $ _____ $7 \times 4 = $ _____

$7 \times 5 = $ _____ $7 \times 6 = $ _____ $7 \times 7 = $ _____ $7 \times 8 = $ _____

$7 \times 9 = $ _____ $7 \times 10 = $ _____ $7 \times 5 = $ _____ $7 \times 6 = $ _____

$7 \times 5 = $ _____ $7 \times 7 = $ _____ $7 \times 5 = $ _____ $7 \times 8 = $ _____

$7 \times 5 = $ _____ $7 \times 9 = $ _____ $7 \times 5 = $ _____ $7 \times 10 = $ _____

$7 \times 6 = $ _____ $7 \times 5 = $ _____ $7 \times 6 = $ _____ $7 \times 7 = $ _____

$7 \times 6 = $ _____ $7 \times 8 = $ _____ $7 \times 6 = $ _____ $7 \times 9 = $ _____

$7 \times 6 = $ _____ $7 \times 7 = $ _____ $7 \times 6 = $ _____ $7 \times 7 = $ _____

$7 \times 8 = $ _____ $7 \times 7 = $ _____ $7 \times 9 = $ _____ $7 \times 7 = $ _____

$7 \times 8 = $ _____ $7 \times 6 = $ _____ $7 \times 8 = $ _____ $7 \times 7 = $ _____

$7 \times 8 = $ _____ $7 \times 9 = $ _____ $7 \times 9 = $ _____ $7 \times 6 = $ _____

$7 \times 9 = $ _____ $7 \times 7 = $ _____ $7 \times 9 = $ _____ $7 \times 8 = $ _____

$7 \times 9 = $ _____ $7 \times 8 = $ _____ $7 \times 6 = $ _____ $7 \times 9 = $ _____

$7 \times 7 = $ _____ $7 \times 9 = $ _____ $7 \times 6 = $ _____ $7 \times 8 = $ _____

$7 \times 9 = $ _____ $7 \times 7 = $ _____ $7 \times 6 = $ _____ $7 \times 8 = $ _____

multiply by 7 (6–10)

EUREKA
MATH™

Name _____ Date _____

1. Each side on a sticky note measures 9 centimeters. What is the area of the sticky note?

2. Stacy tiles the rectangle below using her square pattern blocks.

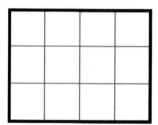

a. Find the area of Stacy's rectangle in square units. Then, draw and label a different rectangle with whole number side lengths that has the same area.

b. Can you draw another rectangle with different whole number side lengths and have the same area? Explain how you know.

3. An artist paints a 4 foot × 16 foot mural on a wall. What is the total area of the mural? Use the break apart and distribute strategy.

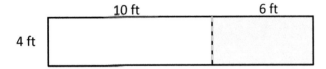

4. Alana tiles the 3 figures below. She says, "I'm making a pattern!"

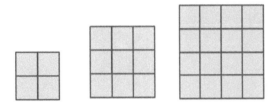

a. Find the area of Alana's 3 figures and explain her pattern.

b. Draw the next 2 figures in Alana's pattern and find their areas.

5. Jermaine glues 3 identical pieces of paper as shown below and makes a square. Find the unknown side length of 1 piece of paper. Then, find the total area of 2 pieces of paper.

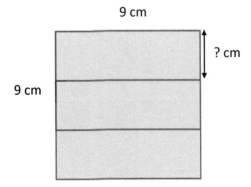

Lesson 12:   Solve word problems involving area.

EUREKA
MATH™

Name _____   Date _____

1. A painting has an area of 63 square inches. One side length is 9 inches. What is the other side length?

9 inches

Area = 63 square inches

2. Judy's mini dollhouse has one floor and measures 4 inches by 16 inches. What is the total area of the dollhouse floor?

Name _____   Date _____

1. A square calendar has sides that are 9 inches long. What is the calendar's area?

2. Each [ ] is 1 square unit. Sienna uses the same square units to draw a 6 × 2 rectangle and says that it has the same area as the rectangle below. Is she correct? Explain why or why not.

3. The surface of an office desk has an area of 15 square feet. Its length is 5 feet. How wide is the office desk?

EUREKA
MATH

4. A rectangular garden has a total area of 48 square yards. Draw and label two possible rectangular gardens with different side lengths that have the same area.

5. Lila makes the pattern below. Find and explain her pattern. Then, draw the *fifth* figure in her pattern.

Lesson 13

Objective: Find areas by decomposing into rectangles or completing composite figures to form rectangles.

Suggested Lesson Structure

■ Fluency Practice   (12 minutes)
▨ Application Problem   (6 minutes)
▢ Concept Development   (32 minutes)
■ Student Debrief   (10 minutes)

   Total Time   **(60 minutes)**

Fluency Practice (12 minutes)

▪ Group Counting **3.OA.1**   (4 minutes)
▪ Find the Common Products **3.OA.7**   (8 minutes)

Group Counting (4 minutes)

Note: Group counting reviews interpreting multiplication as repeated addition.

Instruct students to count forward and backward, occasionally changing the direction of the count.

   ▪ Threes to 30
   ▪ Sixes to 60
   ▪ Eights to 80
   ▪ Nines to 90

Find the Common Products (8 minutes)

Materials: (S) Blank paper

Note: This fluency activity reviews multiplication patterns.

After listing the products of 4 and 8, guide students through the following steps:

   T: Draw a line to match the products that appear in both columns.
   S: (Match 8, 16, 24, 32, and 40.)

Lesson 13:   Find areas by decomposing into rectangles or completing composite
                          figures to form rectangles.
                          ©2015 Great Minds. eureka-math.org
                          G3-M4-TE-B4-1.3.1-01.2016

EUREKA MATH™

T: (Write 2 × 4 = 8, etc., next to each matched product on the left half of the paper.) Write the rest of the number sentences like I did.

S: (Write number sentences.)

T: (Write 8 = 1 × 8, etc., next to each matched product on the right half of the paper.) Write the rest of the number sentences like I did.

S: (Write number sentences.)

T: (Write 2 × 4 = __ × 8.) Say the true number sentence.

S: 2 × 4 = 1 × 8.

T: (Write 2 × 4 = 1 × 8.) Write the remaining equal facts as number sentences.

S: (Write 4 × 4 = 2 × 8, 6 × 4 = 3 × 8, 8 × 4 = 4 × 8, and 10 × 4 = 5 × 8.)

T: Discuss the patterns in your number sentences.

Application Problem (6 minutes)

Anil finds the area of a 5-inch by 17-inch rectangle by breaking it into 2 smaller rectangles. Show one way that he could have solved the problem. What is the area of the rectangle?

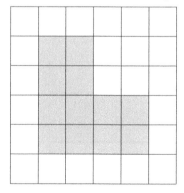

NOTES ON MULTIPLE MEANS OF ENGAGEMENT:

Students who solve the Application Problem quickly may enjoy comparing their solution strategy with others. They may discuss or journal about their reasoning.

Note: This problem reinforces the strategy of breaking apart a larger shape into 2 smaller shapes to find the total area.

Concept Development (32 minutes)

Materials: (S) Personal white board, large grid (Template)

Problem 1: Add using the break apart strategy to find the area of a composite shape.

Distribute one large grid to each student. Draw or project the shaded shape shown to the right.

T: Draw and shade the shape on your grid.

S: (Draw and shade.)

T: How do you find the area of a rectangle?

S: Multiply the side lengths!

Large Grid with Shaded Shape

Lesson 13:   Find areas by decomposing into rectangles or completing composite figures to form rectangles.

161

T: Talk to your partner: Can we find the area of the shaded figure by multiplying side lengths? How do you know?

S: No, because it isn't a rectangle. → We can count the unit squares inside.

T: In the Application Problem, we used the break apart and distribute strategy to find the area of a larger rectangle by breaking it into smaller rectangles. Turn and talk to your partner: How might we use a strategy like that to find the area of the shaded figure?

S: We can break it into a square and a rectangle. → We can break it into three squares.

T: Draw a dotted line to show how to break the shaded figure apart into a square and rectangle.

S: (Draw.)

T: (Model as shown on the right.) What equation tells you the area of the square on the top?

S: 2 × 2 = 4.

T: What equation tells you the area of the rectangle on the bottom?

S: 2 × 4 = 8.

T: How do we use those measurements to find the area of the shaded figure?

MP.7   S: Add them together!

T: What is the sum of 8 and 4?

S: 12.

T: What is the area of the shaded figure?

S: 12 square units!

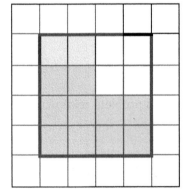

Draw or project the shape shown to the right.

T: We can also find the area of the shaded figure by thinking about a 4 × 4 square with missing units. Turn and talk to your partner: How can we find the shaded area using the unshaded square?

S: The area of the square is 16 square units. → Since the entire square isn't shaded, we need to subtract the 4 square units that are unshaded. → 16 − 4 = 12.

T: There are different strategies of finding the area of a figure. It just depends on how you choose to look at it.

Continue with the following suggested examples:

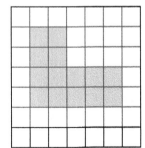 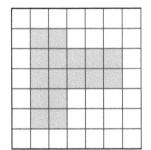 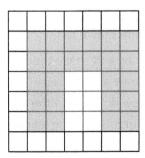

Lesson 13:   Find areas by decomposing into rectangles or completing composite figures to form rectangles.

EUREKA
MATH™

Problem 2: Subtract to find the area of a composite shape.

Draw or project the shape shown to the right.

6 cm

6 cm

4 cm

2 cm

    T: This figure shows a small rectangle cut out of a larger, shaded rectangle. How can we find the area of the shaded figure?

    S: We can break apart the shaded part. → Or we can subtract the unshaded area from the shaded square.

    T: (Shade in the white shape.) We now have a large, shaded square. Write a number sentence to find the area of the large square.

    S: (Write 6 × 6 = 36.)

    T: What is the area of the square?

    S: 36 square centimeters.

    T: (Erase the shading inside the white rectangle.) Beneath the number sentence you just wrote, write a number sentence to find the area for the shape we "cut out."

    S: (Write 2 × 4 = 8.)

    T: What is the area of the white shape?

    S: 8 square centimeters.

    T: The area of the square is 36 square centimeters. We cut out, or took away, 8 square centimeters of shading. Turn and talk to your partner: How can we find the area of the shaded region?

    S: Subtract 8 square centimeters from 36 square centimeters!

    T: Write a number sentence to find the area of the shaded region.

    S: (Write 36 − 8 = 28.)

Continue with the following example:

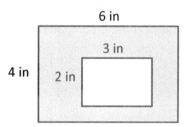

6 in

3 in

4 in

2 in

Problem 3: Subtract to find the area of a composite shape with unknown side lengths.

Draw or project the shape shown to the right.

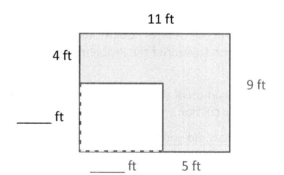

11 ft

4 ft

9 ft

_____ ft

_____ ft   5 ft

    T: This figure also shows a small rectangle cut out of a larger, shaded rectangle, but what is unknown?

    S: The side lengths of the smaller rectangle.

    T: Do we have enough information to find the side lengths of the smaller rectangle?

    S: No, I don't think so. → We know the side lengths of the larger rectangle. → Maybe we can subtract to the find the unknown side lengths.

    Lesson 13:   Find areas by decomposing into rectangles or completing composite figures to form rectangles.

    ©2015 Great Minds. eureka math.org
    G3-M4-TE-B4-1.3.1-01.2016 -

163

T: Opposite sides of a rectangle are equal. Since we know the length of the rectangle is 9 feet, what is the opposite side length?

S: 9 feet.

T: You can then find the unknown lengths by subtracting the known, 4 feet, from the total, 9 feet.

S: The unknown length is 5 feet!

T: Use the same strategy to find the unknown width.

S: (Write 11 − 5 = 6.)

T: What is the unknown width?

S: 6 feet!

T: Can we now find the area of the shaded figure?

S: Yes!

T: With your partner, find the area of the shaded figure.

NOTES ON MULTIPLE MEANS OF ENGAGEMENT:

Extend Problem 3 to students working above grade level. Challenge students to think about a real-life scenario in which this model might be used and to write a matching word problem. Always offer challenges and extensions to learners as alternatives. Here, a student might be given the option of solving one other problem in addition to this extension. Another option would be to direct students to solve the problem that later appears in the Student Debrief.

Problem Set (10 minutes)

Students should do their personal best to complete the Problem Set within the allotted 10 minutes. For some classes, it may be appropriate to modify the assignment by specifying which problems they work on first. Some problems do not specify a method for solving. Students should solve these problems using the RDW approach used for Application Problems.

Student Debrief (10 minutes)

Lesson Objective: Find areas by decomposing into rectangles or completing composite figures to form rectangles.

The Student Debrief is intended to invite reflection and active processing of the total lesson experience.

Invite students to review their solutions for the Problem Set. They should check work by comparing answers with a partner before going over answers as a class. Look for misconceptions or misunderstandings that can be addressed in the Debrief. Guide students in a conversation to debrief the Problem Set and process the lesson.

Any combination of the questions below may be used to lead the discussion.

- How did you break apart the rectangles in Figure 4? Did anyone break apart the rectangles in a different way? (A rectangle of 10 by 2.)

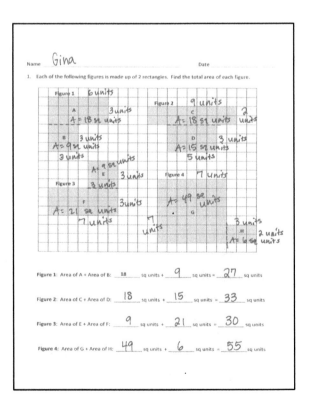

Lesson 13:   Find areas by decomposing into rectangles or completing composite figures to form rectangles.

©2015 Great Minds. eureka-math.org
G3-M4-TE-B4-1.3.1-01.2016

EUREKA MATH™

- In Problem 2, a 4-cm by 3-cm rectangle was cut out of a bigger rectangle. What other measurements could have been cut out to keep the same area for the shaded region?

- How did you find the unknown measurements in Problem 3?

- How were today's strategies examples of using what we know to solve new types of problems?

Exit Ticket (3 minutes)

After the Student Debrief, instruct students to complete the Exit Ticket. A review of their work will help with assessing students' understanding of the concepts that were presented in today's lesson and planning more effectively for future lessons. The questions may be read aloud to the students.

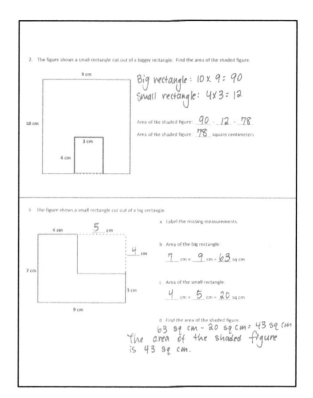

Lesson 13:   Find areas by decomposing into rectangles or completing composite figures to form rectangles.

165

©2015 Great Minds. eureka math.org
G3-M4-TE-B4-1.3.1-01.2016 -

Name _____   Date _____

1.   Each of the following figures is made up of 2 rectangles. Find the total area of each figure.

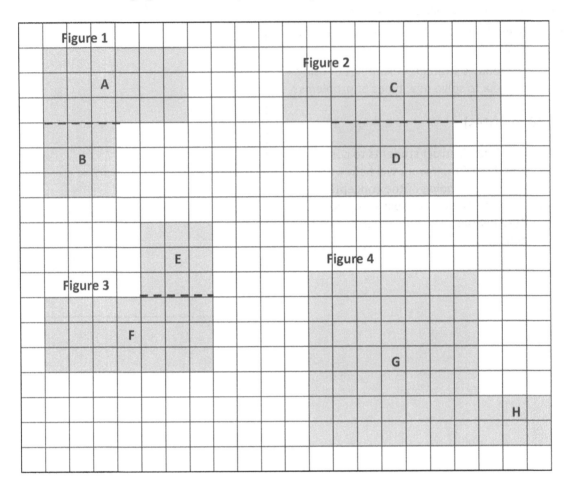

Figure 1: Area of A + Area of B: ___18___ sq units + _____ sq units = _____ sq units

Figure 2: Area of C + Area of D: _____ sq units + _____ sq units = _____ sq units

Figure 3: Area of E + Area of F: _____ sq units + _____ sq units = _____ sq units

Figure 4: Area of G + Area of H: _____ sq units + _____ sq units = _____sq units

Find areas by decomposing into rectangles or completing composite
figures to form rectangles.

EUREKA
MATH

2.   The figure shows a small rectangle cut out of a bigger rectangle. Find the area of the shaded figure.

9 cm

10 cm

3 cm

4 cm

Area of the shaded figure: _____ – _____ = _____

Area of the shaded figure: _____ square centimeters

_____

3.   The figure shows a small rectangle cut out of a big rectangle.

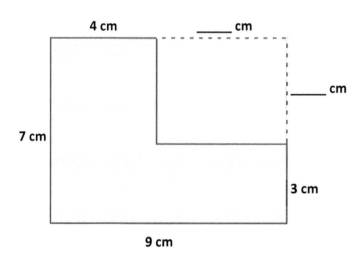

4 cm   _____ cm

7 cm

3 cm

9 cm

a. Label the unknown measurements.

b. Area of the big rectangle:

   _____ cm × _____ cm = _____ sq cm

c. Area of the small rectangle:

   _____ cm × _____ cm = _____ sq cm

d. Find the area of the shaded figure.

Name _____ Date _____

The following figure is made up of 2 rectangles. Find the total area of the figure.

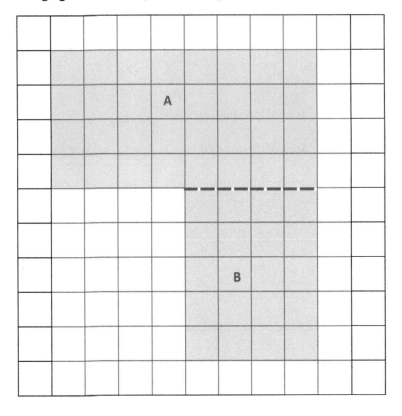

Area of A + Area of B: _____ sq units + _____ sq units = _____ sq units

Lesson 13: Find areas by decomposing into rectangles or completing composite figures to form rectangles.

EUREKA
MATH™

Name _____   Date _____

1.   Each of the following figures is made up of 2 rectangles. Find the total area of each figure.

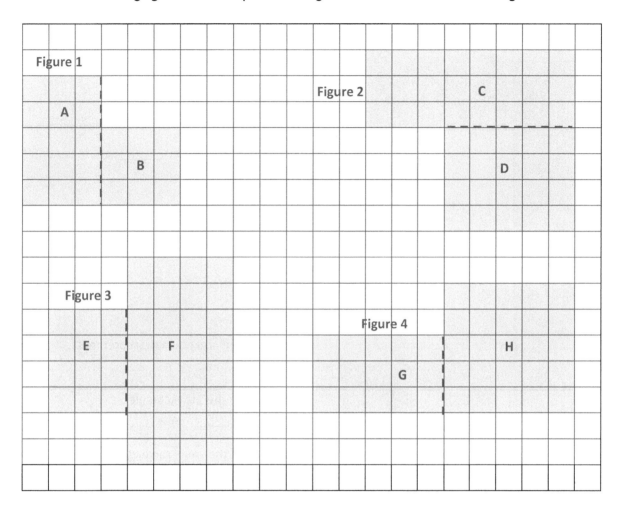

Figure 1: Area of A + Area of B: _____ sq units + _____ sq units = _____ sq units

Figure 2: Area of C + Area of D: _____ sq units + _____ sq units = _____ sq units

Figure 3: Area of E + Area of F: _____ sq units + _____ sq units = _____ sq units

Figure 4: Area of G + Area of H: _____ sq units + _____ sq units = _____ sq units

Lesson 13:   Find areas by decomposing into rectangles or completing composite
figures to form rectangles.

169

©2015 Great Minds. eureka math.org
G3-M4-TE-B4-1.3.1-01.2016 -

2. The figure shows a small rectangle cut out of a big rectangle. Find the area of the shaded figure.

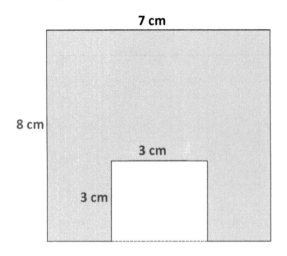

7 cm

8 cm

3 cm

3 cm

Area of the shaded figure: _____ – _____ = _____

Area of the shaded figure: _____ square centimeters

3. The figure shows a small rectangle cut out of a big rectangle.

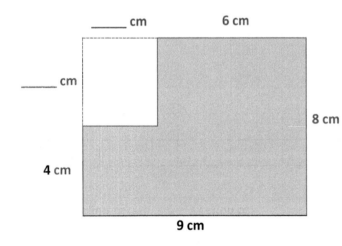

_____ cm

6 cm

_____ cm

8 cm

4 cm

9 cm

a. Label the unknown measurements.

b. Area of the big rectangle:

_____ cm × _____ cm = _____ sq cm

c. Area of the small rectangle:

_____ cm × _____ cm = _____ sq cm

d. Find the area of the shaded figure.

Lesson 13:   Find areas by decomposing into rectangles or completing composite figures to form rectangles.

©2015 Great Minds. eureka-math.org
G3-M4-TE-B4-1.3.1-01.2016

EUREKA MATH

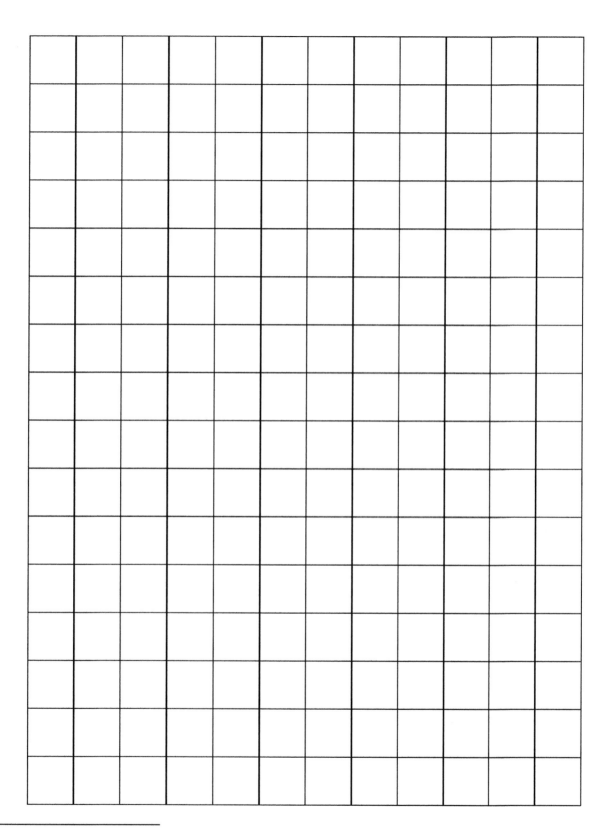

large grid

Lesson 13: Find areas by decomposing into rectangles or completing composite figures to form rectangles.

©2015 Great Minds. eureka math.org
G3-M4-TE-B4-1.3.1-01.2016

Lesson 14

Objective: Find areas by decomposing into rectangles or completing composite figures to form rectangles.

Suggested Lesson Structure

■ Fluency Practice   (15 minutes)
▨ Application Problem   (5 minutes)
▢ Concept Development   (30 minutes)
■ Student Debrief   (10 minutes)

Total Time   **(60 minutes)**

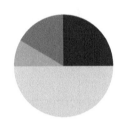

Fluency Practice (15 minutes)

- Group Counting **3.OA.1**   (3 minutes)
- Multiply by 8 **3.OA.7**   (7 minutes)
- Find the Area **3.MD.7**   (5 minutes)

Group Counting (3 minutes)

Note: Group counting reviews interpreting multiplication as repeated addition.

Direct students to count forward and backward, occasionally changing the direction of the count.

- Fours to 40
- Sixes to 60
- Sevens to 70
- Nines to 90

Multiply by 8 (7 minutes)

Materials: (S) Multiply by 8 (6–10) Pattern Sheet

Note: This activity builds fluency with multiplication facts using units of 8. It works toward students knowing from memory all products of two one-digit numbers. See Lesson 2 for the directions for administration of a Multiply-By Pattern Sheet.

T: (Write 6 × 8 = ___.) Let's skip-count up by eights to solve. (Count with fingers to 6 as students count.)

S: 8, 16, 24, 32, 40, 48.

T: Let's skip-count down to find the answer, too. Start at 80. (Count down with fingers as students count.)

S: 80, 72, 64, 56, 48.

T: Let's skip-count up again to find the answer, but this time, start at 40. (Count up with fingers as students count.)

S: 40, 48.

Continue with the following possible sequence: 8 × 8, 7 × 8, and 9 × 8.

T: (Distribute Multiply by 8 (6–10) Pattern Sheet.) Let's practice multiplying by 8. Be sure to work left to right across the page.

Find the Area (5 minutes)

Figures for *Find the Area*

Materials: (S) Personal white board

Note: This fluency activity reviews the relationship between side lengths and area and supports the perception of the composite shapes by moving from part to whole using a grid.

T: (Project the first figure on the right.) On your personal white board, write a number sentence to show the area of the shaded rectangle.

S: (Write 5 × 2 = 10 or 2 × 5 = 10.)

T: Write a number sentence to show the area of the unshaded rectangle.

S: (Write 3 × 2 = 6 or 2 × 3 = 6.)

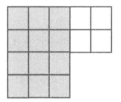

T: (Write ___ sq units + ___ sq units = ___ sq units.) Using the areas of the shaded and unshaded rectangle, write an addition sentence to show the area of the entire figure.

S: (Write 10 sq units + 6 sq units = 16 sq units or 6 sq units + 10 sq units = 16 sq units.)

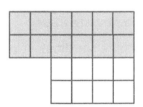

Continue with the other figures.

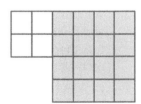

Lesson 14:   Find areas by decomposing into rectangles or completing composite figures to form rectangles.

©2015 Great Minds. eureka math.org
G3-M4-TE-B4-1.3.1-01.2016 -

173

Application Problem (5 minutes)

a. Break apart the shaded figure into 2 rectangles. Then, add to find the area of the shaded figure below.

b. Subtract the area of the unshaded rectangle from the area of the large rectangle to check your answer in Part (a).

MP.7

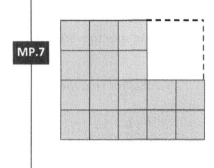

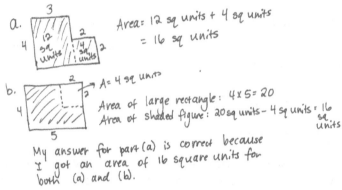

Note: This problem reviews the Lesson 13 concept of finding the area of composite shapes. Students may choose to break apart their rectangles in different ways for Part (a).

Concept Development (30 minutes)

Materials: (S) Personal white board, Problem Set

Problem 1: Choose an appropriate method for finding the area of a composite shape.

Distribute one Problem Set to each student. Project the shape on the right.

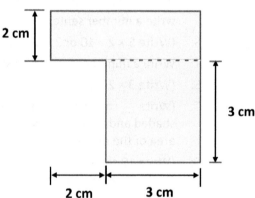

T: What two strategies did we learn yesterday to find the area of a non-rectangular shape?

S: We can break the shape apart into smaller rectangles, and then add the areas of the smaller rectangles together. → Or, we can find the area of the larger rectangle and subtract the area of the unknown part.

T: Look at the figure in Problem 1(a).

T: What is the unknown width?

S: 5 centimeters! → 2 centimeters plus 3 centimeters is 5 centimeters.

T: Label that on your figure. Then, write the equation used to find the area of each of the smaller rectangles.

S: (Record on Problem Set.)

NOTES ON MULTIPLE MEANS OF ENGAGEMENT:

Students working below grade level may benefit from sentence frames to write equations to find the area in Problem 1. Provide the following written support, if necessary:

_____ cm × _____ cm = _____sq cm

_____ cm × _____ cm = _____sq cm

_____ sq cm + _____ sq cm = _____ sq cm

The area is _____ square centimeters.

Lesson 14:   Find areas by decomposing into rectangles or completing composite figures to form rectangles.

©2015 Great Minds. eureka-math.org
G3-M4-TE-B4-1.3.1-01.2016

EUREKA MATH™

T: What is the area of the top rectangle?

S: 10 square centimeters!

T: What is the area of the bottom rectangle?

S: 9 square centimeters!

T: On your Problem Set, write the equation used to find the area of the whole figure. Be sure to answer in a complete sentence!

T: What is the total area of the figure?

S: 19 square centimeters!

Continue with Problem 1(b) from the Problem Set.

Problem 2: Solve a word problem involving the area of non-rectangular shapes.

Write or project the following problem: Fanny has a piece of fabric 8 feet long and 5 feet wide. She cuts out a rectangular piece that measures 3 feet by 2 feet. How many square feet of fabric does Fanny have left?

T: Draw and label Fanny's fabric.

T: How big is the piece that Fanny cuts out?

S: 3 ft by 2 ft.

T: Work with your partner to draw the piece of fabric that Fanny cuts out. Label the measurements of the piece being cut out.
(Note: The 3-ft by 2-ft piece can be taken out of any part of the original rectangle, including at an angle.)

S: (Draw as shown to the right.)

T: What's the best way for us to find the area of the remaining fabric?

S: Find the area of the original piece, then subtract the area of what was cut out.

T: Write an equation to find the area of the original piece of fabric.

S: (Write $8 \times 5 = 40$.)

T: Beneath what you just wrote, write an equation to find the area of the piece of fabric Fanny cuts out.

T: What is the area of the piece that is cut out?

S: 6 square feet!

T: What expression tells us the area of the remaining fabric?

S: $40 - 6$.

T: $40 - 6$ equals...?

S: 34.

T: How much fabric does Fanny have left?

S: 34 square feet!

NOTES ON MULTIPLE MEANS OF ENGAGEMENT:

Adjust the numbers in Problem 2 of the Concept Development to challenge students working above grade level. Or, offer an alternative challenge, such as scripting and recording the steps to find the area of a non-rectangular shape that they can refer to when needed.

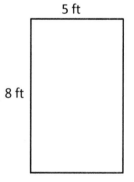

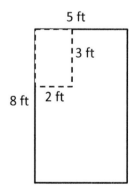

Lesson 14: Find areas by decomposing into rectangles or completing composite figures to form rectangles.

175

©2015 Great Minds. eureka math.org
G3-M4-TE-B4-1.3.1-01.2016 -

Problem Set (10 minutes)

Students should do their personal best to complete the Problem Set within the allotted 10 minutes. For some classes, it may be appropriate to modify the assignment by specifying which problems they work on first. Some problems do not specify a method for solving. Students should solve these problems using the RDW approach used for Application Problems.

Student Debrief (10 minutes)

Lesson Objective: Find areas by decomposing into rectangles or completing composite figures to form rectangles.

The Student Debrief is intended to invite reflection and active processing of the total lesson experience.

Invite students to review their solutions for the Problem Set. They should check work by comparing answers with a partner before going over answers as a class. Look for misconceptions or misunderstandings that can be addressed in the Debrief. Guide students in a conversation to debrief the Problem Set and process the lesson.

Any combination of the questions below may be used to lead the discussion.

- Lead a discussion about the strategy choice for Problems 1(a) and 1(b). Could the strategies have been reversed for these two problems?

- What steps did you need to follow to solve Problem 2? How were you able to find the area of the smaller rectangle?

- Invite students to share their drawings for Problem 3. In what ways are they similar? In what ways are they different?

- Why did Tila and Evan wind up with the same amount of paper in Problem 4? If they both cut their rectangles from the corners of their papers, would they both be able to cut out a 4-cm by 8-cm rectangle with their remaining paper? (Guide students to reason that, although they both have 42 sq cm left and the 4 × 8 rectangle only measures 32 sq cm, only Evan can cut out such a rectangle from his remaining paper.)

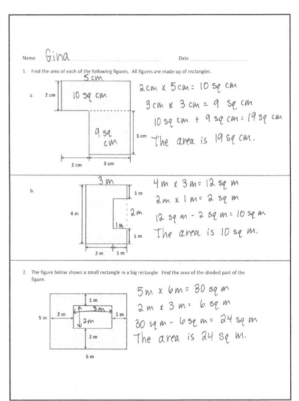

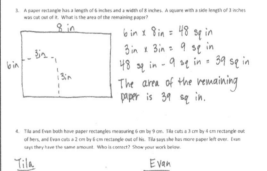

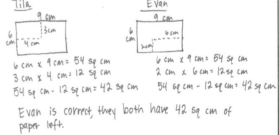

Lesson 14:   Find areas by decomposing into rectangles or completing composite figures to form rectangles.

EUREKA MATH

Exit Ticket (3 minutes)

After the Student Debrief, instruct students to complete the Exit Ticket. A review of their work will help with assessing students' understanding of the concepts that were presented in today's lesson and planning more effectively for future lessons. The questions may be read aloud to the students.

Lesson 14:   Find areas by decomposing into rectangles or completing composite
figures to form rectangles.

©2015 Great Minds. eureka math.org
G3-M4-TE-B4-1.3.1-01.2016 -

177

Multiply.

8 x 1 = _____   8 x 2 = _____   8 x 3 = _____   8 x 4 = _____

8 x 5 = _____   8 x 6 = _____   8 x 7 = _____   8 x 8 = _____

8 x 9 = _____   8 x 10 = _____   8 x 5 = _____   8 x 6 = _____

8 x 5 = _____   8 x 7 = _____   8 x 5 = _____   8 x 8 = _____

8 x 5 = _____   8 x 9 = _____   8 x 5 = _____   8 x 10 = _____

8 x 6 = _____   8 x 5 = _____   8 x 6 = _____   8 x 7 = _____

8 x 6 = _____   8 x 8 = _____   8 x 6 = _____   8 x 9 = _____

8 x 6 = _____   8 x 7 = _____   8 x 6 = _____   8 x 7 = _____

8 x 8 = _____   8 x 7 = _____   8 x 9 = _____   8 x 7 = _____

8 x 8 = _____   8 x 6 = _____   8 x 8 = _____   8 x 7 = _____

8 x 8 = _____   8 x 9 = _____   8 x 9 = _____   8 x 6 = _____

8 x 9 = _____   8 x 7 = _____   8 x 9 = _____   8 x 8 = _____

8 x 9 = _____   8 x 8 = _____   8 x 6 = _____   8 x 9 = _____

8 x 7 = _____   8 x 9 = _____   8 x 6 = _____   8 x 8 = _____

8 x 9 = _____   8 x 7 = _____   8 x 6 = _____   8 x 8 = _____

multiply by 8 (6–10)

Lesson 14: Find areas by decomposing into rectangles or completing composite figures to form rectangles.

EUREKA MATH™

Name _____ Date _____

1. Find the area of each of the following figures. All figures are made up of rectangles.

   a.

   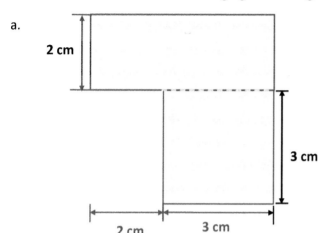

   b.

   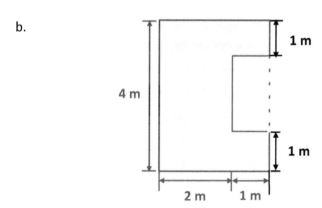

2. The figure below shows a small rectangle in a big rectangle. Find the area of the shaded part of the figure.

   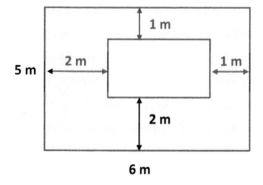

Lesson 14:   Find areas by decomposing into rectangles or completing composite figures to form rectangles.

©2015 Great Minds. eureka math.org
G3-M4-TE-B4-1.3.1-01.2016 -

179

3. A paper rectangle has a length of 6 inches and a width of 8 inches. A square with a side length of 3 inches was cut out of it. What is the area of the remaining paper?

4. Tila and Evan both have paper rectangles measuring 6 cm by 9 cm. Tila cuts a 3 cm by 4 cm rectangle out of hers, and Evan cuts a 2 cm by 6 cm rectangle out of his. Tila says she has more paper left over. Evan says they have the same amount. Who is correct? Show your work below.

Lesson 14: Find areas by decomposing into rectangles or completing composite figures to form rectangles.

©2015 Great Minds. eureka-math.org
G3-M4-TE-B4-1.3.1-01.2016

EUREKA
MATH

Name _____   Date _____

Mary draws an 8 cm by 6 cm rectangle on her grid paper. She shades a square with a side length of 4 cm inside her rectangle. What area of the rectangle is left unshaded?

Lesson 14:   Find areas by decomposing into rectangles or completing composite
figures to form rectangles.

©2015 Great Minds. eureka math.org
G3-M4-TE-B4-1.3.1-01.2016   -

181

Name _____   Date _____

1. Find the area of each of the following figures. All figures are made up of rectangles.

    a.

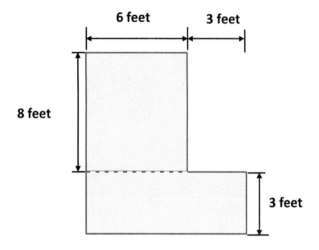

    b.

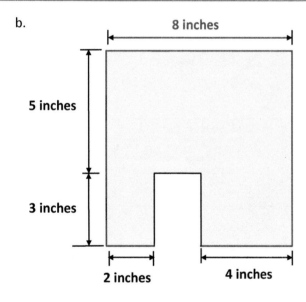

Lesson 14:   Find areas by decomposing into rectangles or completing composite
figures to form rectangles.

©2015 Great Minds. eureka-math.org
G3-M4-TE-B4-1.3.1-01.2016

**EUREKA
MATH**

2. The figure below shows a small rectangle cut out of a big rectangle.

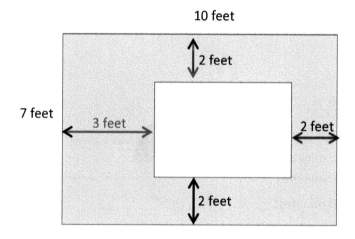

10 feet

2 feet

7 feet

3 feet

2 feet

2 feet

a. Label the side lengths of the unshaded region.

b. Find the area of the shaded region.

Lesson 14:   Find areas by decomposing into rectangles or completing composite
figures to form rectangles.

©2015 Great Minds. eureka math.org
G3-M4-TE-B4-1.3.1-01.2016

183

Lesson 15

Objective: Apply knowledge of area to determine areas of rooms in a given floor plan.

Suggested Lesson Structure

■ Fluency Practice   (15 minutes)
▢ Concept Development   (35 minutes)
■ Student Debrief   (10 minutes)
  Total Time   **(60 minutes)**

Fluency Practice (15 minutes)

- Group Counting **3.OA.1**   (3 minutes)
- Multiply by 9 **3.OA.7**   (7 minutes)
- Find the Area **3.MD.7**   (5 minutes)

Group Counting (3 minutes)

Note: Group counting reviews interpreting multiplication as repeated addition.

Instruct students to count forward and backward, occasionally changing the direction of the count.

- Threes to 30
- Sixes to 60
- Sevens to 70
- Eights to 80

Multiply by 9 (7 minutes)

Materials: (S) Multiply by 9 (1–5) Pattern Sheet

Note: This activity builds fluency with multiplication facts using units of 9. It works toward students knowing all products of two one-digit numbers from memory. See Lesson 2 for the directions for administration of a Multiply-By Pattern Sheet.

T: (Write 5 × 9 = ____.) Let's skip-count by nines to find the answer. (Count with fingers to 5 as students count.)

S: 9, 18, 27, 36, 45. (Record on the board as students count.)

Lesson 15:   Apply knowledge of area to determine areas of rooms in a given floor plan.

©2015 Great Minds. eureka-math.org
G3-M4-TE-B4-1.3.1-01.2016

T: (Circle 45 and write 5 × 9 = 45 above it. Write 3 × 9 = ____.) Let's skip-count up by nines again. (Count with fingers to 3 as students count.)

S: 9, 18, 27.

T: Let's see how we can skip-count down to find the answer, too. Start at 45 with 5 fingers, 1 for each 9. (Count down with your fingers as students say numbers.)

S: 45 (5 fingers), 36 (4 fingers), 27 (3 fingers).

Repeat the process for 4 × 9.

T: (Distribute Multiply by 9 (1–5) Pattern Sheet.) Let's practice multiplying by 9. Be sure to work left to right across the page.

Find the Area (5 minutes)

Materials: (S) Personal white board

Note: This fluency activity reviews the relationship between side lengths and area; additionally, it supports the perception of the composite shape by moving from part to whole using a grid.

T: (Project the figure on the right.) On your personal white board, write a number sentence to show the area of the shaded rectangle.

S: (Write 4 × 2 = 8 or 2 × 4 = 8.)

T: Write a number sentence to show the area of the unshaded rectangle.

S: (Write 3 × 2 = 6 or 2 × 3 = 6.)

T: (Write ___ sq units + ___ sq units = ___ sq units.) Using the areas of the shaded and unshaded rectangles, write an addition sentence to show the area of the entire figure.

S: (Write 8 sq units + 6 sq units = 14 sq units or 6 sq units + 8 sq units = 14 sq units.)

Continue with the figures below:

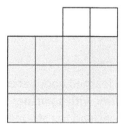 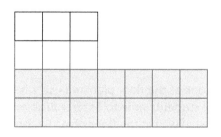 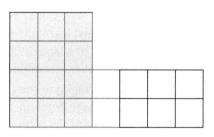

EUREKA
MATH™

Lesson 15:   Apply knowledge of area to determine areas of rooms in a given floor plan.

©2015 Great Minds. eureka math.org
G3-M4-TE-B4-1.3.1-01.2016 -

185

Concept Development (35 minutes)

Materials: (T) Chart paper labeled *Strategies We Can Use to Find Area* (S) Problem Set, ruler

T: For the next two days, you are going to be architects. Today you are going to use a floor plan that your clients designed to find the area in square centimeters of each room in the house. Look at the floor plan. What will you need to do before you can find the areas?

S: We need to find the side lengths of each room. → We need to know the lengths and widths of the rooms.

T: Use your ruler to measure the side lengths of Bedroom 1 in centimeters. What is the length?

S: 5 centimeters.

T: What is the width?

S: 12 centimeters.

T: Write an expression to show how to find the area of Bedroom 1.

S: (Write 5 × 12.)

T: (Write *Multiply Side Lengths* on a chart labeled *Strategies We Can Use to Find Area*.) What strategy can you use to find the area since this fact is so large?

S: The break apart and distribute strategy!

T: (Add the strategy to the chart.) What about the rooms that aren't rectangles, how will you find their areas?

S: We can find the areas of smaller rectangles and add them together to get the area of a room that isn't rectangular. → Yes, that's the break apart and add strategy we just learned. → Or, we might be able to find the area of a large rectangle and then subtract the area of a smaller rectangle.

T: (Add the strategies to the chart.) Look at the floor plan and use what we've learned about area to help you answer Problem 1. (Allow students time to answer Problem 1.) Work with a partner to find the areas of the rooms and the hallway in the floor plan. Record the areas and the strategy you use to find each area in the chart in Problem 2.

A NOTE TO THE TEACHER:

This lesson is designed to be completed in two days. For early finishers, please refer to the optional activities suggested in Lesson 16.

NOTES ON MULTIPLE MEANS OF ACTION AND EXPRESSION:

Some students may benefit from a review of how to use a ruler to measure. Have them try the following:

- Place the zero end of the ruler against the line to be measured.
- Make sure the zero tick mark is lined up against the beginning of the side length.
- Read the last number on the ruler that is by the end of the side length.

To make measuring easier, try the tips below:

- Darken the lines to be measured.
- Outline the lines with glue to make a tactile model.
- Provide large print rulers.
- Give the option of using centimeter blocks to measure.

186   **Lesson 15:**   Apply knowledge of area to determine areas of rooms in a given floor plan.

©2015 Great Minds. eureka-math.org
G3-M4-TE-B4-1.3.1-01.2016

Problem Set (20 minutes)

Students should do their personal best to complete the Problem Set within the allotted 20 minutes. For some classes, it may be appropriate to modify the assignment by specifying which problems they work on first. Some problems do not specify a method for solving. Students should solve these problems using the RDW approach used for Application Problems.

Student Debrief (10 minutes)

Lesson Objective: Apply knowledge of area to determine areas of rooms in a given floor plan.

The Student Debrief is intended to invite reflection and active processing of the total lesson experience.

Invite students to review their solutions for the Problem Set. They should check work by comparing answers with a partner before going over answers as a class. Look for misconceptions or misunderstandings that can be addressed in the Debrief. Guide students in a conversation to debrief the Problem Set and process the lesson.

Any combination of the questions below may be used to lead the discussion.

- Explain to a partner your choice for the prediction you made in Problem 1. What have you learned about area that helped you make your prediction?

- What strategy did you use to find the area of the living room? Is there more than one way to break apart the living room into smaller rectangles? Explain two different ways to a partner.

NOTES ON MULTIPLE MEANS OF ACTION AND EXPRESSION:

To ease the task of constructing a response for Problems 3–5 of the Problem Set, allow English language learners and others to discuss their reasoning before writing. Discussions can be in first languages, if beneficial. Also provide English language learners with sentence frames, such as those given below.

- The _____ has the biggest area. My prediction was right/wrong because_____.

- There are/are not enough tiles because _____.

Name Gina   Date

1. Make a prediction: Which room looks like it has the biggest area?

   I think the living room has the biggest area.

2. Record the areas and show the strategy you used to find each area.

Room	Area	Strategy
Bedroom 1	60 sq cm	$5 \times 12 = 5 \times (10 + 2)$ $= (5 \times 10) + (5 \times 2)$ $= 50 + 10 = 60$
Bedroom 2	56 sq cm	$8 \times 7 = 56$
Kitchen	42 sq cm	$6 \times 7 = 42$
Hallway	24 sq cm	$3 \times 8 = 24$
Bathroom	25 sq cm	$5 \times 5 = 25$
Dining Room	28 sq cm	$4 \times 7 = 28$
Living Room	88 sq cm	$(6 \times 10) + (4 \times 7)$ $= 60 + 28$ $= 88$

Lesson 15:   Apply knowledge of area to determine areas of rooms in a given floor plan.

©2015 Great Minds. eureka math.org
G3-M4-TE-B4-1.3.1-01.2016

187

Exit Ticket (3 minutes)

After the Student Debrief, instruct students to complete the Exit Ticket. A review of their work will help with assessing students' understanding of the concepts that were presented in today's lesson and planning more effectively for future lessons. The questions may be read aloud to the students.

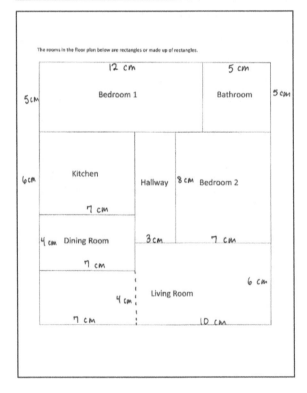

3. Which room has the biggest area? Was your prediction right? Why or why not?

The living room has the biggest area. Yes, my prediction was right because when you add the areas of the small rectangles in the living room, they add up to more than any other room.

4. Find the side lengths of the house without using your ruler to measure them, and explain the process you used.

Side lengths: __19__ centimeters and __17__ centimeters

I added the side lengths of the rooms to find the side lengths of the house, like this:
12 cm + 5 cm = 17 cm
5 cm + 6 cm + 4 cm + 4 cm = 19 cm

5. What is the area of the whole floor plan? How do you know?

Area = __323__ square centimeters

I found the area of the house by adding the areas of the rooms.

42 sq cm + 60 sq cm + 56 sq cm + 24 sq cm + 25 sq cm + 28 sq cm + 88 sq cm = 323 sq cm

The rooms in the floor plan below are rectangles or made up of rectangles.

12 cm — Bedroom 1   5 cm — Bathroom   5 cm   5 cm
6 cm   Kitchen   Hallway   8 cm Bedroom 2
7 cm   3 cm   7 cm
4 cm Dining Room   6 cm
7 cm   4 cm   Living Room
7 cm   10 cm

Lesson 15:   Apply knowledge of area to determine areas of rooms in a given floor plan.

©2015 Great Minds. eureka-math.org
G3-M4-TE-B4-1.3.1-01.2016

EUREKA MATH™

Multiply.

9 x 1 = _____   9 x 2 = _____   9 x 3 = _____   9 x 4 = _____

9 x 5 = _____   9 x 1 = _____   9 x 2 = _____   9 x 1 = _____

9 x 3 = _____   9 x 1 = _____   9 x 4 = _____   9 x 1 = _____

9 x 5 = _____   9 x 1 = _____   9 x 2 = _____   9 x 3 = _____

9 x 2 = _____   9 x 4 = _____   9 x 2 = _____   9 x 5 = _____

9 x 2 = _____   9 x 1 = _____   9 x 2 = _____   9 x 3 = _____

9 x 1 = _____   9 x 3 = _____   9 x 2 = _____   9 x 3 = _____

9 x 4 = _____   9 x 3 = _____   9 x 5 = _____   9 x 3 = _____

9 x 4 = _____   9 x 1 = _____   9 x 4 = _____   9 x 2 = _____

9 x 4 = _____   9 x 3 = _____   9 x 4 = _____   9 x 5 = _____

9 x 4 = _____   9 x 5 = _____   9 x 1 = _____   9 x 5 = _____

9 x 2 = _____   9 x 5 = _____   9 x 3 = _____   9 x 5 = _____

9 x 4 = _____   9 x 2 = _____   9 x 4 = _____   9 x 3 = _____

9 x 5 = _____   9 x 3 = _____   9 x 2 = _____   9 x 4 = _____

9 x 3 = _____   9 x 5 = _____   9 x 2 = _____   9 x 4 = _____

multiply by 9 (1–5)

Lesson 15:   Apply knowledge of area to determine areas of rooms in a given floor plan.

©2015 Great Minds. eureka math.org
G3-M4-TE-B4-1.3.1-01.2016   -

Name _____ Date _____

1. Make a prediction: Which room looks like it has the biggest area?

2. Record the areas and show the strategy you used to find each area.

Room	Area	Strategy
Bedroom 1	_____ sq cm	
Bedroom 2	_____ sq cm	
Kitchen	_____ sq cm	
Hallway	_____ sq cm	
Bathroom	_____ sq cm	
Dining Room	_____ sq cm	
Living Room	_____ sq cm	

Lesson 15:   Apply knowledge of area to determine areas of rooms in a given floor plan.

©2015 Great Minds. eureka-math.org
G3-M4-TE-B4-1.3.1-01.2016

EUREKA
MATH™

3. Which room has the biggest area? Was your prediction right? Why or why not?

4. Find the side lengths of the house without using your ruler to measure them, and explain the process you used.

   Side lengths: _____ centimeters and _____ centimeters

5. What is the area of the whole floor plan? How do you know?

   Area = _____ square centimeters

Lesson 15:   Apply knowledge of area to determine areas of rooms in a given floor plan.

©2015 Great Minds. eureka math.org
G3-M4-TE-B4-1.3.1-01.2016 -

191

The rooms in the floor plan below are rectangles or made up of rectangles.

Bedroom 1	Bathroom

Kitchen	Hallway	Bedroom 2
Dining Room		

Living Room

Lesson 15:   Apply knowledge of area to determine areas of rooms in a given floor plan.

©2015 Great Minds. eureka-math.org
G3-M4-TE-B4-1.3.1-01.2016

EUREKA
MATH

Name _____ Date _____

Jack uses grid paper to create a floor plan of his room. Label the unknown measurements, and find the area of the items listed below.

Name	Equations	Total Area
a. Jack's Room		_____ square units
b. Bed		_____ square units
c. Table		_____ square units
d. Dresser		_____ square units
e. Desk		_____ square units

EUREKA
MATH™

Lesson 15: Apply knowledge of area to determine areas of rooms in a given floor plan.

©2015 Great Minds. eureka math.org
G3-M4-TE-B4-1.3.1-01.2016 -

193

Name _____   Date _____

Use a ruler to measure the side lengths of each numbered room in centimeters. Then, find the area. Use the measurements below to match, and label the rooms with the correct areas.

Kitchen: 45 square centimeters   Living Room: 63 square centimeters

Porch: 34 square centimeters   Bedroom: 56 square centimeters

Bathroom: 24 square centimeters   Hallway: 12 square centimeters

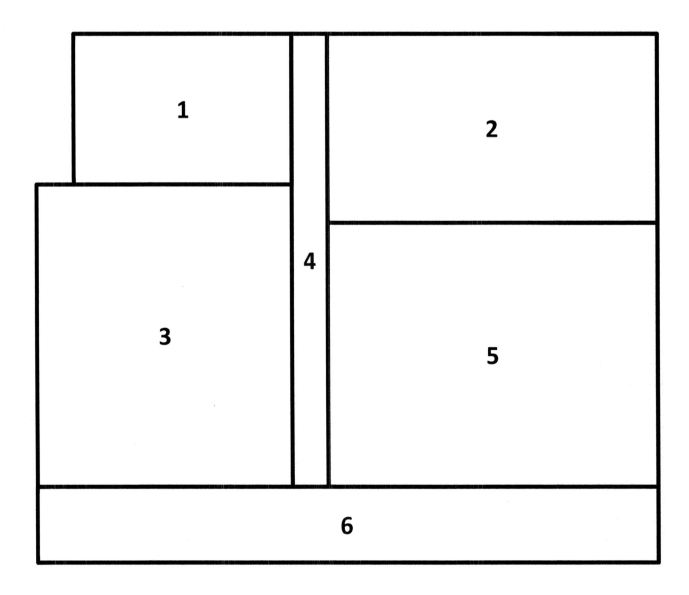

EUREKA
MATH™

Lesson 16

Objective: Apply knowledge of area to determine areas of rooms in a given floor plan.

Suggested Lesson Structure

■ Fluency Practice   (15 minutes)
▨ Concept Development   (35 minutes)
■ Student Debrief   (10 minutes)

Total Time   **(60 minutes)**

Fluency Practice (15 minutes)

- Group Counting **3.OA.1**   (3 minutes)
- Multiply by 9 **3.OA.7**   (7 minutes)
- Find the Area **3.MD.7**   (5 minutes)

Group Counting (3 minutes)

Note: Group counting reviews interpreting multiplication as repeated addition.

Instruct students to count forward and backward, occasionally changing the direction of the count.

- Sixes to 60
- Sevens to 70
- Eights to 80

Multiply by 9 (7 minutes)

Materials: (S) Multiply by 9 (6–10) Pattern Sheet

Note: This activity builds fluency with multiplication facts using units of 9. It works toward students knowing all products of two one-digit numbers from memory. See Lesson 2 for the directions for administration of a Multiply-By Pattern Sheet.

T: (Write $6 \times 9 =$ ____.) Let's skip-count up by nine to solve. (Count with fingers to 6 as students count.)

S: 9, 18, 27, 36, 45, 54.

T: Let's skip-count down to find the answer, too. Start at 90. (Count down with fingers as students count.)

S: 90, 81, 72, 63, 54.

T: Let's skip-count up again to find the answer, but this time start at 45. (Count up with fingers as students count.)

S: 45, 54.

Continue with the following possible sequence: 8 × 9, 7 × 9, and 9 × 9.

T: (Distribute Multiply by 9 (6–10) Pattern Sheet.) Let's practice multiplying by 9. Be sure to work left to right across the page.

Find the Area (5 minutes)

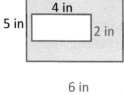

Materials: (S) Personal white board

Note: This fluency activity reviews Lesson 14.

T: (Project the first figure on the right.) Find the areas of the large rectangle and the unshaded rectangle. Then, subtract to find the area of the shaded figure. (Write *Area = ____ square inches*.)

S: (Work and write *Area = 27 square inches*.)

Continue with other figures.

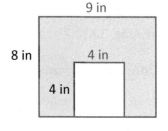

Concept Development (35 minutes)

Materials: (S) Lesson 15 Problem Set, ruler

T: Today you will continue to find the area of each room in the house in square centimeters.

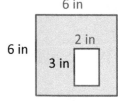

If students finish early, feel free to use one or both of the following options.

Option 1: Create a floor plan with different side lengths for given areas.

Materials: (S) Centimeter grid (Lesson 3 Template 1), construction paper, glue

Students work with a partner to create a floor plan with the areas of the rooms that they found. The task is for students to find new side lengths for each room. Students should use their answers from the Problem Set to ensure that they find different side lengths with the same area. After they find new side lengths, they mark each room on centimeter grid paper and then cut out the rooms. They use these centimeter grids to fit the rooms together to make their floor plan. They glue their final arrangement of rooms onto a piece of construction paper. Allow students a few minutes to do a gallery walk of the completed floor plans.

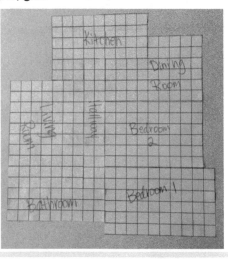

Lesson 16:   Apply knowledge of area to determine areas of rooms in a given floor plan.

©2015 Great Minds. eureka-math.org
G3-M4-TE-B4-1.3.1-01.2016

EUREKA
MATH™

Option 2: Review strategies to find new side lengths of given areas.

Materials: (S) Lesson 16 Problem Set

T: Yesterday you found the areas of the rooms in a floor plan that your clients designed. They like the area of each room, but they want to change the way the rooms look. Your job today is to create rooms with the same areas, but with different side lengths. Are you up for the challenge, architects?

S: Yes!

T: Look at the Problem Set. What is the area of the hallway?

S: 24 square centimeters.

T: What are possible side lengths you can have for the hallway and still have the same area?

S: 3 and 8. → 1 and 24. → 2 and 12. → 6 and 4.

T: Talk to a partner: Which of these choices was used in the floor plan?

S: 8 and 3. → The numbers are just switched.

T: So, when you redesign the floor plan today, be sure you don't use that combination!

Student Debrief (10 minutes)

Lesson Objective: Apply knowledge of area to determine areas of rooms in a given floor plan.

The Student Debrief is intended to invite reflection and active processing of the total lesson experience.

Invite students to review their solutions for the Problem Set. They should check work by comparing answers with a partner before going over answers as a class. Look for misconceptions or misunderstandings that can be addressed in the Debrief. Guide students in a conversation to debrief the Problem Set and process the lesson.

Any combination of the questions below may be used to lead the discussion.

- Explain to a partner how you found the side lengths of the whole house without using your ruler to measure.

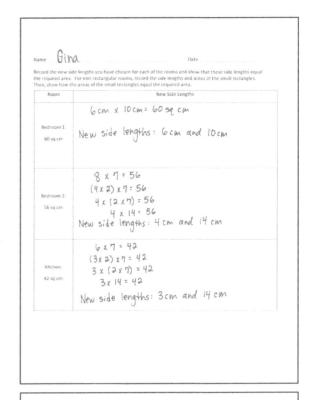

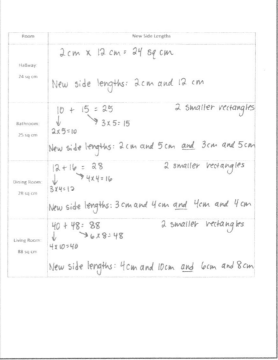

- Can you multiply the side lengths of the house to find the area of the house? Why or why not? How did you find the area of the whole house?

- Do we usually measure rooms in centimeters? What unit might each centimeter in this picture represent on a real house? (Yards, feet, or meters.)

Exit Ticket (3 minutes)

After the Student Debrief, instruct students to complete the Exit Ticket. A review of their work will help with assessing students' understanding of the concepts that were presented in today's lesson and planning more effectively for future lessons. The questions may be read aloud to the students.

Lesson 16:   Apply knowledge of area to determine areas of rooms in a given floor
plan.

©2015 Great Minds. eureka-math.org
G3-M4-TE-B4-1.3.1-01.2016

Multiply.

9 x 1 = _____   9 x 2 = _____   9 x 3 = _____   9 x 4 = _____

9 x 5 = _____   9 x 6 = _____   9 x 7 = _____   9 x 8 = _____

9 x 9 = _____   9 x 10 = _____   9 x 5 = _____   9 x 6 = _____

9 x 5 = _____   9 x 7 = _____   9 x 5 = _____   9 x 8 = _____

9 x 5 = _____   9 x 9 = _____   9 x 5 = _____   9 x 10 = _____

9 x 6 = _____   9 x 5 = _____   9 x 6 = _____   9 x 7 = _____

9 x 6 = _____   9 x 8 = _____   9 x 6 = _____   9 x 9 = _____

9 x 6 = _____   9 x 7 = _____   9 x 6 = _____   9 x 7 = _____

9 x 8 = _____   9 x 7 = _____   9 x 9 = _____   9 x 7 = _____

9 x 8 = _____   9 x 6 = _____   9 x 8 = _____   9 x 7 = _____

9 x 8 = _____   9 x 9 = _____   9 x 9 = _____   9 x 6 = _____

9 x 9 = _____   9 x 7 = _____   9 x 9 = _____   9 x 8 = _____

9 x 9 = _____   9 x 8 = _____   9 x 6 = _____   9 x 9 = _____

9 x 7 = _____   9 x 9 = _____   9 x 6 = _____   9 x 8 = _____

9 x 9 = _____   9 x 7 = _____   9 x 6 = _____   9 x 8 = _____

multiply by 9 (6–10)

Lesson 16:   Apply knowledge of area to determine areas of rooms in a given floor plan.

199

©2015 Great Minds. eureka math.org
G3-M4-TE-B4-1.3.1-01.2016 -

Name _____ Date _____

Record the new side lengths you have chosen for each of the rooms and show that these side lengths equal the required area. For non-rectangular rooms, record the side lengths and areas of the small rectangles. Then, show how the areas of the small rectangles equal the required area.

Room	New Side Lengths
Bedroom 1: 60 sq cm	
Bedroom 2: 56 sq cm	
Kitchen: 42 sq cm	

Lesson 16: Apply knowledge of area to determine areas of rooms in a given floor plan.

©2015 Great Minds. eureka-math.org
G3-M4-TE-B4-1.3.1-01.2016

EUREKA
MATH™

Room	New Side Lengths
Hallway: 24 sq cm	
Bathroom: 25 sq cm	
Dining Room: 28 sq cm	
Living Room: 88 sq cm	

Lesson 16:   Apply knowledge of area to determine areas of rooms in a given floor
plan.

©2015 Great Minds. eureka math.org
G3-M4-TE-B4-1.3.1-01.2016 -

201

Name _____ Date _____

Find the area of the shaded figure. Then, draw and label a rectangle with the same area.

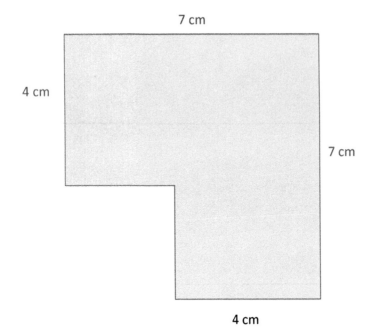

7 cm

4 cm

7 cm

4 cm

Apply knowledge of area to determine areas of rooms in a given floor
plan.

EUREKA
MATH

Name _____ Date _____

Jeremy plans and designs his own dream playground on grid paper. His new playground will cover a total area of 100 square units. The chart shows how much space he gives for each piece of equipment, or area. Use the information in the chart to draw and label a possible way Jeremy can plan his playground.

Basketball court	10 square units
Jungle gym	9 square units
Slide	6 square units
Soccer area	24 square units

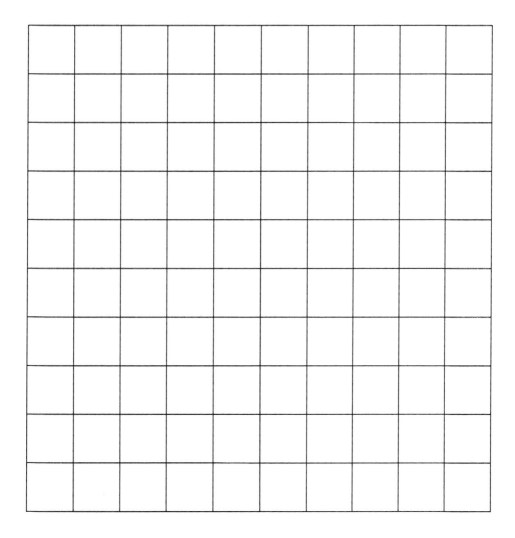

Lesson 16: Apply knowledge of area to determine areas of rooms in a given floor plan.

©2015 Great Minds. eureka math.org
G3-M4-TE-B4-1.3.1-01.2016 -

203

Name _____   Date _____

1. Sarah says the rectangle on the left has the same area as the sum of the two on the right. Pam says they do not have the same areas. Who is correct? Explain using numbers, pictures, and words.

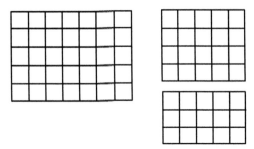

2. Draw three different arrays that you could make with 36 square inch tiles. Label the side lengths on each of your arrays. Write multiplication sentences for each array to prove that the area of each array is 36 square inches.

EUREKA
MATH

3.   Mr. and Mrs. Jackson are buying a new house. They are deciding between the two floor plans below.

House A

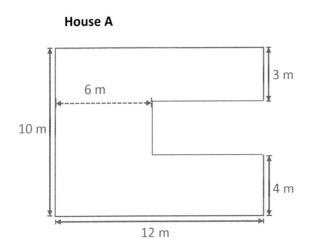

House B

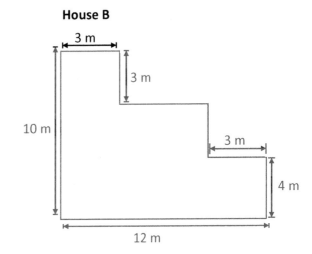

Which floor plan has the greater area? Show how you found your answer on the drawings above. Show your calculations below.

4. Superior Elementary School uses the design below for their swimming pool. Shapes A, B, and C are rectangles.

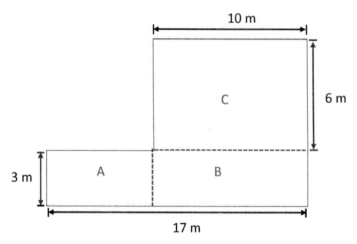

a. Label the side lengths of Rectangles A and B on the drawing.

b. Find the area of each rectangle.

c. Find the area of the entire pool. Explain how you found the area of the pool.

©2015 Great Minds. eureka-math.org
G3-M4-TE-B4-1.3.1-01.2016

EUREKA
MATH™

Geometric measurement: understand concepts of area and relate area to multiplication and to addition.

3.MD.5   Recognize area as an attribute of plane figures and understand concepts of area measurement.

       a.   A square with side length 1 unit, called "a unit square," is said to have "one square unit" of area, and can be used to measure area.

       b.   A plane figure which can be covered without gaps or overlaps by n unit squares is said to have an area of n square units.

3.MD.6   Measure areas by counting unit squares (square cm, square m, square in, square ft, and improvised units).

3.MD.7   Relate area to the operations of multiplication and addition.

       a.   Find the area of a rectangle with whole-number side lengths by tiling it, and show that the area is the same as would be found by multiplying the side lengths.

       b.   Multiply side lengths to find areas of rectangles with whole-number side lengths in the context of solving real world and mathematical problems, and represent whole-number products as rectangular areas in mathematical reasoning.

       c.   Use tiling to show in a concrete case that the area of a rectangle with whole-number side lengths a and $b + c$ is the sum of $a \times b$ and $a \times c$. Use area models to represent the distributive property in mathematical reasoning.

       d.   Recognize area as additive. Find areas of rectilinear figures by decomposing them into non-overlapping rectangles and adding the areas of the non-overlapping parts, applying this technique to solve real world problems.

Evaluating Student Learning Outcomes

A Progression Toward Mastery is provided to describe steps that illuminate the gradually increasing understandings that students develop on their way to proficiency. In this chart, this progress is presented from left (Step 1) to right (Step 4). The learning goal for students is to achieve Step 4 mastery. These steps are meant to help teachers and students identify and celebrate what the students CAN do now and what they need to work on next.

©2015 Great Minds. eureka math.org
G3-M4-TE-B4-1.3.1-01.2016 -

A Progression Toward Mastery

Assessment Task Item and Standards Assessed	STEP 1 Little evidence of reasoning without a correct answer. (1 Point)	STEP 2 Evidence of some reasoning without a correct answer or with a partially correct answer in a multi-step question. (2 Points)	STEP 3 Evidence of some reasoning with a correct answer or evidence of solid reasoning with an incorrect answer. (3 Points)	STEP 4 Evidence of solid reasoning with a correct answer. (4 Points)
1 **3.MD.7c** **3.MD.7d**	Response demonstrates little or no evidence of reasoning without a correct answer.	Student identifies that Sarah is correct, demonstrating evidence of limited reasoning to support the answer.	Student identifies that Sarah is correct. Response shows evidence of accurate reasoning to support the answer using at least one representation.	Student identifies that Sarah is correct. Explanation shows evidence of solid reasoning using multiple representations.
2 **3.MD.5b** **3.MD.6** **3.MD.7a** **3.MD.7b**	Student attempts but is unable to draw any correct arrays with labels. Multiplication sentences are not shown.	Student correctly draws and labels one array. Side lengths are labeled without units. A multiplication sentence is shown.	Student correctly draws and labels two different arrays. Side lengths are labeled in inches. Multiplication sentences are shown for those two arrays.	Student correctly draws and labels three different arrays. Side lengths are labeled in inches. Possible arrays are as follows: ▪ 1×36 ▪ 2×18 ▪ 3×12 ▪ 4×9 ▪ 6×6 Correct multiplication sentences are shown for each array drawn.
3 **3.MD.7d** **3.MD.7b**	Response demonstrates little or no evidence of reasoning without a correct answer.	Student miscalculates one area. Student may identify that House A has the greater area with limited reasoning.	Response demonstrates correct calculations and area. Student identifies that House A has the greater area.	Student demonstrates correct area calculations with answers: ▪ House A = 102 sq meters ▪ House B = 84 sq meters Explanation identifies that House A has the greater area. Response provides evidence of solid reasoning.

EUREKA
MATH™

A Progression Toward Mastery

| 4

3.MD.5
3.MD.7b
3.MD.7d | Attempts but is unable to answer any part of the question correctly. | Student does the following:

a. Labels length and width correctly but without units.

b. Calculates at least two areas correctly.

c. May miscalculate the total area. | Student answers Parts (a) and (b) correctly but may miscalculate the total area. | Student correctly:

a. Labels length and width of rectangles A and B, including the following units:

  ■ A = 3 m × 7 m
  ■ B = 3 m × 10 m

b. Calculates the area of each rectangle as follows:

  ■ A = 21 sq meters
  ■ B = 30 sq meters
  ■ C = 60 sq meters

c. Calculates the total area as 111 sq meters. |

Name **Gina**   Date

1. Sarah says the rectangle on the left has the same area as the sum of the two on the right. Pam says they do not have the same areas. Who is correct? Explain using numbers, pictures, and words.

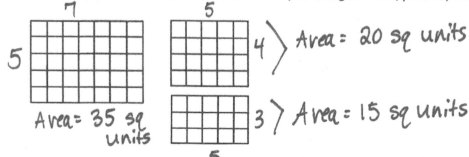

Sarah is correct. The two on the right add up to 35 sq units, which is the area of the one on the left.

2. Draw three different arrays that you could make with 36 square inch tiles. Label the side lengths on each of your arrays. Write multiplication sentences for each array to prove that the area of each array is 36 square inches.

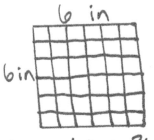

6 in × 6 in = 36 sq in

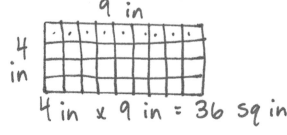

4 in × 9 in = 36 sq in

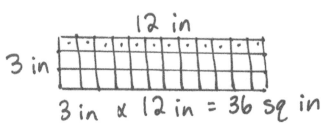

3 in × 12 in = 36 sq in

EUREKA
MATH™

3. Mr. and Mrs. Jackson are buying a new house. They are deciding between the two floor plans below.

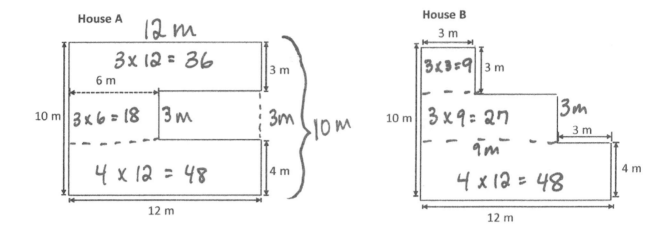

House A

12 m

3 x 12 = 36

6 m

10 m   3 x 6 = 18   3 m   3m   } 10 m

3m

4 x 12 = 48   4 m

12 m

House B

3 m

3×3=9   3 m

10 m   3 x 9 = 27   3m

9m   3 m

4 x 12 = 48   4 m

12 m

Which floor plan has the greater area? Show how you found your answer on the drawings above. Show your calculations below.

House A:

36 + 18 + 48

40 + 14

54 + 48

52 + 50

= 102

Area: 102 sq m

House B:

9 + 27 + 48

6 + 30

36 + 48

34 + 50

= 84

Area: 84 sq m

House A has the greater area because it is 102 square meters and House B is only 84 square meters.

4. Superior Elementary School uses the design below for their swimming pool. Shapes A, B, and C are rectangles.

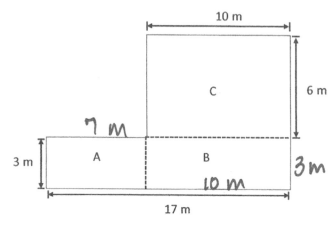

a. Label the side lengths of Rectangles A and B on the drawing.

b. Find the area of each rectangle.

A: 3 m × 7 m = 21 sq m

B: 3 m × 10 m = 30 sq m

C: 6 m × 10 m = 60 sq m

c. Find the area of the entire pool. Explain how you found the area of the pool.

21 sq m + 30 sq m + 90 sq m   The area of the pool is 111 square meters. I found the area by adding the areas of all 3 parts.

21 sq m + 90 sq m

111 sq m

EUREKA MATH

Eureka Math
Grade 3
Module 4

Special thanks go to the Gordon A. Cain Center and to the Department of Mathematics at Louisiana State University for their support in the development of *Eureka Math*.

For a free *Eureka Math* Teacher
Resource Pack, Parent Tip
Sheets, and more please
visit www.Eureka.tools

Answer Key

GRADE 3 • MODULE 4

Multiplication and Area

©2015 Great Minds. eureka math.org
G3-M4-TE-B4-1.3.1-01.2016 -

Lesson 1

Problem Set

1. Lines drawn to show 6 triangles each inside Shapes A and B

2. Lines drawn to show 3 rhombuses each inside Shapes A and B

3. Lines drawn to show 2 trapezoids each inside Shapes A and B

4. As pattern blocks get bigger, the number of blocks it takes to cover the same shape gets smaller.

5. Lines drawn to show 6 squares inside rectangle

6. No, because you can't have gaps or overlaps when measuring area

Exit Ticket

Yes, both rectangles have an area of 12 square units.

Homework

1. a. 12

    b. 6

    c. 4; because $12 \div 3 = 4$

2. a. 12

    b. 12; because 12 squares fit inside of it

3. A, because it has an area of 18 square units

Module 4:   Multiplication and Area

EUREKA
MATH™

Lesson 2

Pattern Sheet

4	8	12	16
20	24	28	32
36	40	24	28
24	32	24	36
24	40	24	28
24	28	32	28
36	28	40	28
32	24	32	28
32	36	32	40
32	36	24	36
28	36	32	36
40	36	40	24
40	28	40	32
40	36	40	24
32	40	28	36

Problem Set

1. Rectangle A: 2 rows of 6 square inches; 12 sq in

   Rectangle B: 3 rows of 4 square inches; 12 sq in

   Rectangle C: 4 rows of 3 square inches; 12 sq in

2. Rectangle A: 2 rows of 6 square centimeters; 12 sq cm

   Rectangle B: 3 rows of 4 square centimeters; 12 sq cm

   Rectangle C: 4 rows of 3 square centimeters; 12 sq cm

3. Answers will vary.

4. Yes, 6 square units inside of each rectangle

5. 8 square units; rectangle with an area of 8 square units drawn

Exit Ticket

1. 16 square units; rectangle with an area of 16 square units drawn

2. No, square inches are larger than square centimeters.

Homework

1. a. 12; rectangle circled

   b. 9

   c. 12; rectangle circled

   d. 12; rectangle circled

   e. 5

   f. 8

2. No, rectangle with 8 square units; rectangle with 6 square units

3. 16 square units; rectangle with an area of 16 square units drawn

EUREKA
MATH™

Lesson 3

Problem Set

1. A: 6

    B: 2 square units

    C: 12 square units

    D: 20 square units

2. a. 6 square units

    b. 9 square units

    c. 16 square units

    d. 12 square units

3. a. Answers will vary.

    b. Answers will vary.

4. Answers will vary.

Exit Ticket

1. 12 square units; rectangle with an area of 12 square units drawn

2. No, this rectangle's area is 16 square units.

Homework

1. A: 5

    B: 15 square units

    C: 12 square units

    D: 20 square units

2. a. 9 square units

    b. 24 square units

    c. 8 square units

    d. 18 square units

3. A: 10; rectangle with an area of 10 square units drawn

    B: 9 square units; rectangle with an area of 9 square units drawn

    C: 12 square units; rectangle with an area of 12 square units drawn

Lesson 4

Problem Set

1. 2 cm by 7 cm marked and connected; 14 sq cm

2. 3 in by 2 in marked and connected; 6 sq in

3. 3 cm by 4 cm labeled; 12 sq cm

4. Both are correct; explanations will vary.

5. Square-inch tiles; explanations will vary.

6. Explanations will vary.

Exit Ticket

a. 2 cm by 3 cm; 6 sq cm

b. 2 cm by 6 cm; 12 sq cm

c. 1 in by 5 in; 5 sq in

Homework

1. 8 sq cm

2. 4 cm by 5 cm labeled; 20 sq cm

3. 2 in by 7 in labeled; 14 sq in

4. Both are correct; explanations will vary.

5. 2 in; 4 in; 8 sq in; explanations will vary.

Module 4:   Multiplication and Area

©2015 Great Minds. eureka-math.org
G3-M4-TE-B4-1.3.1-01.2016

EUREKA
MATH™

Lesson 5

Problem Set

1. a. 6 cm; 6

    b. 20; tiles drawn; 4, 5, 20

    c. 3 cm; tiles drawn; 6, 3, 18

    d. 8 cm; tiles drawn; 3, 8, 24

    e. 4 cm; tiles drawn; 5, 4, 20

    f. 9; tiles drawn; 3, 3, 9

2. 5 in, 7 in; answers will vary.

3. Two rectangular arrays drawn, multiplication sentences written for each

4. a. 8; answers will vary.

    b. No; answers will vary.

Exit Ticket

Rectangle with 7 rows of 4 tiles drawn; labeled 7 cm and 4 cm; multiplication sentence written

Homework

1. a. 6 cm; 6

    b. 4 cm; tiles drawn; 6, 4, 24

    c. 3 cm; tiles drawn; 5, 3, 15

    d. 5 cm; tiles drawn; 3, 5, 15

2. 9; answers will vary.

3. a. 9; answers will vary.

    b. Yes; answers will vary.

    c. Yes; explanations will vary.

Lesson 6

Problem Set

1. a. Lines drawn to find 5 cm by 6 cm; matched to fifth completed array; 5, 6, 30

    b. Lines drawn to find 3 cm by 7 cm; matched to sixth completed array; 3, 7, 21

    c. Lines drawn to find 5 cm by 3 cm; matched to first completed array; 5, 3, 15

    d. Lines drawn to find 4 cm by 5 cm; matched to second completed array; 4, 5, 20

    e. Lines drawn to find 2 cm by 6 cm; matched to third completed array; 2, 6, 12

    f. Lines drawn to find 4 cm by 3 cm; matched to fourth completed array; 4, 3, 12

2. No; explanations may vary.

3. 90

4. 30; explanations may vary.

Exit Ticket

80

Homework

1. a. Lines drawn to find 6 cm by 6 cm; matched to fifth completed array; 6, 6, 36

    b. Lines drawn to find 3 cm by 8 cm; matched to sixth completed array; 3, 8, 24

    c. Lines drawn to find 3 cm by 6 cm; matched to first completed array; 3, 6, 18

    d. Lines drawn to find 5 cm by 5 cm; matched to second completed array; 5, 5, 25

    e. Lines drawn to find 2 cm by 8 cm; matched to third completed array; 2, 8, 16

    f. Lines drawn to find 4 cm by 3 cm; matched to fourth completed array; 4, 3, 12

2. Yes; explanations may vary.

3. 90

4. 30; explanations may vary.

EUREKA
MATH™

Lesson 7

Problem Set

1. a. Grid lines drawn inside rectangle; side lengths labeled; 3, 4, 12

   b. Grid lines drawn inside rectangle; side lengths labeled; 5, 4, 20

   c. Grid lines drawn inside rectangle; side lengths labeled; 2, 7, 14

   d. Grid lines drawn inside rectangle; side lengths labeled; 7, 4, 28

   e. Grid lines drawn inside rectangle; side lengths labeled; 1, 3, 3

   f. Grid lines drawn inside rectangle; side lengths labeled; 4, 2, 8

2. a. Side lengths labeled as 9 feet and 11 feet

   b. Grid lines drawn inside rectangle

   c. 99

3. No; explanations will vary.

4. a. Answers will vary.

   b. 24

Exit Ticket

1. Grid lines drawn inside rectangle; side lengths labeled; 42

2. Gia, square inches are larger than square centimeters.

Homework

1. a. 6; answer provided; 2, 6

   b. 10; side lengths labeled; 2, 5, 10

   c. 12; side lengths labeled; 3, 4, 12

   d. 16; side lengths labeled; 4, 4, 16

2. a. 7 by 4 rectangle drawn on grid; 28 square units

   b. Side lengths labeled; $7 \times 4 = 28$

3. Gregory, square inches are larger than square centimeters.

Lesson 8

Pattern Sheet

6	12	18	24
30	36	42	48
54	60	30	36
30	42	30	48
30	54	30	60
36	30	36	42
36	48	36	54
36	42	36	42
48	42	54	42
48	36	48	42
48	54	54	36
54	42	54	48
54	48	36	54
42	54	36	48
54	42	36	48

Problem Set

1. a. 28; 4, 7, 28
   b. 56; 8, 7, 56
   c. 36; 6, 6, 36
2. a. 8; 9, 8, 72; 72, 9, 8
   b. 5; 3, 5, 15; 15, 3, 5
   c. 7; 7, 4, 28; 28, 4, 7

3. Answers will vary.
4. 54 sq cm; explanations will vary.
5. No; explanations will vary.
6. 4 in; explanations will vary.

Exit Ticket

1. 27; 3, 9, 27
2. 9; 6, 9, 54; 54, 6, 9

Module 4:   Multiplication and Area

EUREKA MATH

Homework

1. a. 24; 3, 8, 24

    b. 48; 6, 8, 48

    c. 16; 4, 4, 16

    d. 28; 4, 7, 28

2. a. 8; 3, 8, 24; 24, 3, 8

    b. 4; 4, 9, 36; 36, 9, 4

3. Answers will vary.

4. 36 sq cm; explanations will vary.

5. 3 in; explanations will vary.

©2015 Great Minds. eureka math.org
G3-M4-TE-B4-1.3.1-01.2016 -

Lesson 9

Problem Set

1. a. 2 rectangles drawn; 5 cm, 10 cm labeled
    b. 5 cm × 10 cm = 50 sq cm
    c. 50 sq cm + 50 sq cm = 100 sq cm
2. a. Rectangle drawn; 5 cm, 20 cm labeled
    b. 100 sq cm

3. a. 4, 6; 4, 7; 24 sq units; 28 sq units
    b. Rectangle drawn; 4, 13
    c. Rahema is right; explanations will vary.
4. No; explanations will vary.

Exit Ticket

1. 6, 6; 6, 3
2. 6 × 6 = 36, 36 sq units; 6 × 3 = 18, 18 sq units
3. 54 sq units; answers will vary.

Homework

1. a. Line drawn to show two 4 by 8 rectangles or two 8 by 4 rectangles; 1 rectangle shaded
    b. 4, 8; 4, 8 or 8, 4; 8, 4
    c. 4 × 8 + 4 × 8 = 64,
        8 × 4 + 8 × 4 = 64, or 8 × 8 = 64

2. a. Rectangle drawn; 4, 16
    b. 64 sq units
    c. Yes; answers will vary.

Module 4:   Multiplication and Area

EUREKA MATH

Lesson 10

Problem Set

1. a. 35, 21; 56; 56

    b. 10; 10; 10; 40; 48; 48

    c. 10, 3; 10; 10; 60, 18; 78; 78

    d. 8, 10, 2; 10, 2; 10, 2; 80, 16; 96; 96

2. Answers will vary.

3. 75 sq units; answers will vary.

Exit Ticket

1. 8, 5, 2; 5, 2; 5, 2; 40, 16; 56; 56

2. 9, 10, 3; 10, 3; 9, 10, 9, 3; 90, 27; 117; 117

Homework

1. a. 40, 32; 72; 72

    b. 10; 10; 10; 50; 60; 60

    c. 10, 3; 10; 10; 70, 21; 91; 91

    d. 9, 10, 2; 10, 2; 10, 2; 90, 18; 108; 108

2. Answers will vary.

3. Rectangle shaded; 64 sq units; answers will vary.

Lesson 11

Problem Set

1. a. 6, 48; 48

    b. 48, 48; 48

    c. 24; 2, 24; 48; 48

    d. 12; 4, 12; 48; 48

    e. 16, 3; 16, 3; 48; 48

2. Yes; answers will vary.

3. Answers will vary.

4. a. 72 sq cm

    b. 8, 9; 72; 72; yes; answers will vary.

    c. Answers will vary.

Exit Ticket

1. 64 sq cm

2. 4, 16; 4, 16; 64; 64

Homework

1. a. 9, 36; 36

    b. 36; 36

    c. 18; 2, 18; 36; 36

    d. 12, 3; 12, 3; 36; 36

    e. 6, 6; 6, 6; 36; 36

2. Yes, answers will vary.

3. a. 48 sq cm

    b. 8, 6; 48; 48; yes; answers will vary.

    c. Answers will vary.

EUREKA
MATH™

Lesson 12

Pattern Sheet

7	14	21	28
35	42	49	56
63	70	35	42
35	49	35	56
35	63	35	70
42	35	42	49
42	56	42	63
42	49	42	49
56	49	63	49
56	42	56	49
56	63	63	42
63	49	63	56
63	56	42	63
49	63	42	56
63	49	42	56

Problem Set

1. 81 sq cm

2. a. 12 sq units; answers will vary.

   b. Yes; answers will vary.

3. 64 sq ft

4. a. 4 sq units, 9 sq units, 16 sq units; explanations will vary.

   b. 5 by 5 and 6 by 6 rectangles drawn; 25 sq units, 36 sq units

5. 3 cm; 54 sq cm

Exit Ticket

1. 7 in

2. 64 sq in

Homework

1. 81 sq in

2. Yes; answers will vary.

3. 3 ft

4. 2 rectangles drawn; answers will vary.

5. 5 by 2 rectangle drawn; explanations will vary.

228

Module 4:   Multiplication and Area

©2015 Great Minds. eureka-math.org
G3-M4-TE-B4-1.3.1-01.2016

EUREKA
MATH™

Lesson 13

Problem Set

1.   9, 27; 18, 15, 33; 9, 21, 30; answers will vary, 55

2.   90, 12, 78; 78

3.   a.   5, 4

      b.   7, 9, 63

      c.   4, 5, 20

      d.   43 sq cm

Exit Ticket

32, 20, 52

Homework

1.   15, 9, 24; 24, 20, 44; 12, 32, 44; 15, 25, 40

2.   56, 9, 47; 47

3.   a.   4, 3

      b.   8, 9, 72

      c.   4, 3, 12

      d.   60 sq cm

Lesson 14

Pattern Sheet

8	16	24	32
40	48	56	64
72	80	40	48
40	56	40	64
40	72	40	80
48	40	48	56
48	64	48	72
48	56	48	56
64	56	72	56
64	48	64	56
64	72	72	48
72	56	72	64
72	64	48	72
56	72	48	64
72	56	48	64

Problem Set

1. a. 19 sq cm
   b. 10 sq m
2. 24 sq m

3. 39 sq in
4. Evan; explanations will vary.

Exit Ticket

32 sq cm

Homework

1. a. 75 sq ft
   b. 58 sq in

2. a. 3 ft, 5 ft
   b. 55 sq ft

EUREKA MATH™

Lesson 15

Pattern Sheet

9	18	27	36
45	9	18	9
27	9	36	9
45	9	18	27
18	36	18	45
18	9	18	27
9	27	18	27
36	27	45	27
36	9	36	18
36	27	36	45
36	45	9	45
18	45	27	45
36	18	36	27
45	27	18	36
27	45	18	36

Problem Set

1. Answers will vary.

2. 60; 56; 42; 24; 25; 28; 88; strategies will vary.

3. Living room; yes or no; answers will vary.

4. 19, 17; answers will vary.

5. 323; answers will vary.

Exit Ticket

Missing side lengths labeled

    a.   Equations will vary, 300

    b.   Equations will vary, 60

    c.   Equations will vary, 9

    d.   Equations will vary, 24

    e.   Equations will vary, 21

Homework

1.   4, 6; bathroom; 24 sq cm

2.   5, 9; kitchen; 45 sq cm

3.   8, 7; bedroom; 56 sq cm

4.   12, 1; hallway; 12 sq cm

5.   7, 9; living room; 63 sq cm

6.   2, 17; porch; 34 sq cm

EUREKA
MATH™

Lesson 16

Pattern Sheet

9	18	27	36
45	54	63	72
81	90	45	54
45	63	45	72
45	81	45	90
54	45	54	63
54	72	54	81
54	63	54	63
72	63	81	63
72	54	72	63
72	81	81	54
81	63	81	72
81	72	54	81
63	81	54	72
81	63	54	72

Problem Set

Answers will vary.

Exit Ticket

40 sq cm; rectangle drawn, side lengths labeled

Homework

Drawings will vary.

This page intentionally left blank

This page intentionally left blank

This page intentionally left blank